STUDY GUIDE

D0534524

BETTE KREUZ
University of Michigan, Dearborn

General, Organic, and Biological

CHEMISTRY

CONCISE | PRACTICAL | INTEGRATED

FROST
DEAL

Second Edition

PEARSON

Boston Columbus Indianapolis New York San Francisco Upper Saddle River
Amsterdam Cape Town Dubai London Madrid Milan Munich Paris Montréal Toronto
Delhi Mexico City São Paulo Sydney Hong Kong Seoul Singapore Taipei Tokyo

Editor in Chief: Adam Jaworski

Executive Editor: Jeanne Zalesky

Senior Marketing Manager: Jonathan Cottrell

Project Editor: Jessica Moro

Editorial Assistant: Lisa Tarabokjia

Managing Editor, Chemistry and Geosciences: Gina M. Cheselka

Production Project Manager: Connie M. Long

Operations Specialist: Jeffrey Sargent

Cover Photo Credit: Mark Conklin/V & W/imagequestmarine.com

Credits and acknowledgments borrowed from other sources and reproduced, with permission, in this textbook appear on the appropriate page within the text.

Copyright © 2014, 2011 by Pearson Education, Inc. All rights reserved. Manufactured in the United States of America. This publication is protected by Copyright, and permission should be obtained from the publisher prior to any prohibited reproduction, storage in a retrieval system, or transmission in any form or by any means: electronic, mechanical, photocopying, recording, or likewise. To obtain permission(s) to use material from this work, please submit a written request to Pearson Education, Inc., Permissions Department, 1 Lake Street, Department 1G, Upper Saddle River, NJ 07458.

Many of the designations used by manufacturers and sellers to distinguish their products are claimed as trademarks. Where those designations appear in this book, and the publisher was aware of a trademark claim, the designations have been printed in initial caps or all caps.

PEARSON

www.pearsonhighered.com

4 5 6 7 8 9 10—V069—17 16 15

ISBN-10: 0-321-83426-7; ISBN-13: 978-0-321-83426-3

Chemistry Basics—
Matter and Measurement

1.1 Classifying Matter: Mixture or Pure Substance

Learning Objectives

Upon completion of this material, a student should be able to do the following:
A. Define the following key terms:

mixture   element
homogeneous mixture   atom
heterogeneous mixture   compound
pure substance

B. Classify matter as a pure substance of a mixture.
C. Classify mixtures as homogeneous mixture or heterogeneous.
D. Classify pure substances as elements or compounds.

1.2 Elements, Compounds, and the Periodic Table

Learning Objectives

Upon completion of this material, a student should be able to do the following:
A. Define the following key terms:

periodic table of elements   period
chemical symbol   chemical formula
group

B. Distinguish between groups and periods.
C. Locate metals and nonmetals on the periodic table.
D. Identify the number of elements in a chemical formula.

1.3 Math Counts

Learning Objectives

Upon completion of this material, a student should be able to do the following:
A. Define the following key terms:

metric system   equivalent units
Système International (SI) units   conversion factor
kilogram   dimensional analysis
liter   significant figures
meter   coefficient

B. Convert between metric units.
C. Apply the appropriate number of significant figures to a calculation.
D. Convert numbers to scientific notation.
E. Convert numbers and fraction to percent.

Copyright © 2014 Pearson Education, Inc.

1.4 Matter: The "Stuff" of Matter

Learning Objectives

Upon completion of this material, a student should be able to do the following:

A. Define the following key terms:

mass	**law of conservation of energy**
volume	**calorie**
density	**Calorie**
specific gravity	**specific heat capacity**
temperature	**state of matter**
energy	**solid**
potential energy	**liquid**
kinetic energy	**gas**
mass number	

B. Define mass and its measurement.
C. Define volume and its measurement.
D. Calculate and solve problems using density.
E. Convert temperature among the three temperature scales.
F. Distinguish kinetic and potential energy.
G. Convert between energy units.
H. Compare specific heat values of various materials.
I. Contrast the properties of solids, liquids, and gases.

1.5 Measuring Matter

Learning Objectives

Upon completion of this material, a student should be able to do the following:

A. Define the following key terms:

accuracy	**precision**

B. Distinguish between accuracy and precision.
C. Convert SI and U.S. units.
D. Apply conversion factors, units, and percent to measurements in health.

1.6 How Matter Changes

Learning Objectives

Upon completion of this material, a student should be able to do the following:

A. Define the following key terms:

physical change	**reactant**
chemical reaction	**product**
chemical equation	**law of conservation of mass**

B. Distinguish physical changes and chemical reactions.
C. Balance a given chemical equation.

Copyright © 2014 Pearson Education, Inc.

Practice Test for Chapter 1

1. Which of the following is a pure substance?
   A. copper wire
   B. a wooden baseball bat
   C. iced tea
   D. clean air

2. Which of the following statements about heterogeneous mixtures is **false**?
   A. The individual components are distinguishable.
   B. Components of the mixture can be separated by their physical properties.
   C. They are composed of two or more elements that are chemically joined together.
   D. No two samples contain the same substances in the same amount.

3. Which of the following is a mixture?
   A. muddy water
   B. table salt, NaCl
   C. oxygen, O_2
   D. aluminum foil

4. Which of the following is a compound?
   A. CO
   B. Cu
   C. Al
   D. Sn

5. What are the atomic symbols of silver and iron?
   A. Ag and Fe
   B. Au and Pb
   C. Si and I
   D. S and Ir

6. Which of the following elements is an alkaline earth metal?
   A. potassium
   B. strontium
   C. neon
   D. iodine

7. What main group element is located in Group 4A of Period 3 on the Periodic Table?
   A. Si
   B. Ga
   C. Ge
   D. In

Copyright © 2014 Pearson Education, Inc.

8. Which of the following compounds contains six oxygen atoms?
   A. $C_6H_{12}O_6$
   B. $Mg(NO_3)_2$
   C. $Ca_3(PO_3)_2$
   D. all of the above

9. Which of the following compounds contains the element sodium?
   A. SO_2
   B. $NaNO_3$
   C. K_3PO_4
   D. Sm_2O_3

10. Length is measured in which of the following units?
    A. cc
    B. mg
    C. km
    D. cal

11. Mass is measured in which of the following units?
    A. kL
    B. ns
    C. J
    D. mg

12. Which of the following is larger than 10 mg?
    A. 0.0000001 g
    B. 0.0001 kg
    C. 1 cg
    D. 10,000 µg

13. How many mL are in 0.0452 L?
    A. 45.2 mL
    B. 0.0000452 mL
    C. 4.52 mL
    D. 22.6 mL

14. In one second, light travels 29,979,245,800 cm. How many kilometers is this?
    A. 299,792.458 km
    B. 2,997,924,580 km
    C. 2,997,924,580,000,000 km
    D. 299,792,458,000 km

15. Which of the following has more than two significant figures?
    A. 20
    B. 1.20
    C. 430000
    D. 0.015

Copyright © 2014 Pearson Education, Inc.

16. Perform the addition $53,511 + 4,200$ and report your answer to the correct number of significant figures.
    A. 58,000
    B. 57,700
    C. 57,711
    D. 49,300

17. Do the following calculation and report your answer to the correct number of significant figures.
    $$0.00040 / 0.000000602 = ?$$
    A. 664.5
    B. 665
    C. 660
    D. 670

18. Express the following number in scientific notation.
    $$0.0000416$$
    A. 4.16×10^{-5}
    B. 4.16×10^{5}
    C. 4.16×10^{-6}
    D. 416×10^{-3}

19. Express the following number in decimal format.
    $$2.67 \times 10^{4}$$
    A. 267
    B. 2,670
    C. 26,700
    D. 0.000267

20. Express 1/4 and 0.075 as percents, respectively.
    A. 4.0%, 7.5%
    B. 25%, 7.5%
    C. 25%, 75%
    D. 40%, 75%

21. What is 8.00% of 555?
    A. 1.44
    B. 444
    C. 69.4
    D. 44.4

22. Mercury has a density of 13.6 g/mL. What is the mass in kilograms of 25.9 mL of mercury?
    A. 352 kg
    B. 1.90 kg
    C. 0.352 kg
    D. 0.00190 kg

Copyright © 2014 Pearson Education, Inc.

23. A Sahara desert summer could be as hot as 133 °F. What is this temperature on the Kelvin scale?
   A. 329 K
   B. 133 K
   C. −217 K
   D. 13 K

24. A medium hardboiled egg is listed as containing 71 Cal.   Calculate the number of joules in the egg.
   A. 7.1×10^4 J
   B. 3.0×10^5 J
   C. 1.5×10^4 J
   D. 7.1×10^{-2} J

25.   Consider the data given in this table taken from the text.

TABLE 1.4 Specific Heats of Various Substances

Substance	Specific Heat (cal/g °C)
Water (liquid)	1.00
Human body	0.83
Paraffin wax	0.60
Wood, soft	0.34
Wood, hard	0.29
Air	0.24
Aluminum	0.21
Table salt	0.21
Brick	0.20
Stainless steel	0.12
Iron	0.11
Copper	0.092
Silver	0.056
Gold	0.031

For 1-g samples of each of the following substances, which will experience the smallest temperature change when 100 J of energy is added?
   A. paraffin wax
   B. air
   C. stainless steel
   D. gold

Copyright © 2014 Pearson Education, Inc.

26. A liquid has _____.
    A. neither definite shape nor volume
    B. both definite shape and volume
    C. a definite shape but no definite volume
    D. a definite volume but no definite shape

27. The mass of an unknown was repeatedly measured and the results are given below.

$$42.9 \text{ g}, \quad 42.8 \text{ g}, \quad 43.1 \text{ g}, \, 43.0 \text{ g}$$

If the actual mass of the unknown is 50.0 g, this series of measurements is:
A. accurate but not precise.
B. precise but not accurate.
C. both accurate and precise.
D. neither accurate nor precise.

28. Which of the following is a physical change?
    A. rusting iron
    B. digesting food
    C. freezing water
    D. burning propane

29. Which of the following is a chemical change?
    A. boiling water
    B. making ammonia from nitrogen and hydrogen
    C. melting lead
    D. mixing salt with sand

30. How many reactant oxygen atoms will be present in the balanced equation for the following reaction?

$$C_2H_6O(l) \; + \; O_2(g) \rightarrow CO_2(g) + H_2O(g)$$

    A. 3
    B. 5
    C. 6
    D. 7

Copyright © 2014 Pearson Education, Inc.

Answers

1. A   2. C   3. A   4. A   5. A   6. B   7. A   8. D   9. B   10. C

11. D   12. B   13. A   14. A   15. B   16. B   17. C   18. A   19. C   20. B

21. D   22. C   23. A   24. B   25. A   26. D   27. B   28. C   29. B   30. D

Copyright © 2014 Pearson Education, Inc.

Chapter 1 – Solutions to Odd-Numbered Problems

Practice Problems

1.1   a. heterogeneous   b. homogeneous   c. homogeneous   d. heterogeneous

1.3   a. pure substance   b. mixture   c. pure substance

1.5   If it is found on the periodic table, it is an element.

    a. element   b. compound   c. element   d. compound

1.7   (answers in boldface)

Name	Elemental Symbol	Group	Period	Metal or Nonmetal
Fluorine	**F**	**7A**	**2**	**Nonmetal**
Lithium	Li	1A	2	**Metal**
Chlorine	Cl	**7A**	**3**	**Nonmetal**
Carbon	**C**	4A	2	Nonmetal

1.9   a. 1 sodium, 1 nitrogen, 2 oxygens
    b. 12 carbons, 22 hydrogens, 11 oxygens
    c. 1 calcium, 1 carbon, 3 oxygens

1.11   a. 2   b. 4   c. 5   d. 3

1.13   a. 2000   b. 54.5   c. 224,000   d. 132

1.15

$$325 \text{ mg} \times \frac{1 \text{g}}{1000 \text{ mg}} = 0.325 \text{ g}$$

1.17   a. 2.03×10^8 b. 1.24×10^1 c. 2.78×10^{-8}

1.19   a. 0.0000156   b. 280,000   c. 0.090

1.21   a. 25%   b. 38%   c. 66%

1.23   a. 0.045   b. 0.130   c. 0.66   d. 0.78

1.25   a. 20   b. 90   c. 38   d. 133

1.27   a. less   b. more   c. less

1.29   39 g

1.31   446 mL

1.33   104.5 °F

1.35   Before eating, the food is mostly potential energy, during eating it is a mixture of potential and kinetic energy as it is broken down, and after eating the food gets transformed into mostly kinetic energy to carry out the functions of the body.

1.37   86 cal

1.39   The warehouse made of pine. The pine has a higher specific heat, so it will take more heat energy to raise the temperature of the wood versus the brick. This will result in a slower transfer of the heat to the contents of the warehouse.

1.41   a. gas   b. solid

1.43   a. neither   b. both   c. good precision

1.45   150 g

1.47   100 mg/day; 50 mg/dose

1.49   71.4%

1.51   a. chemical reaction
    b. physical change
    c. chemical reaction

Copyright © 2014 Pearson Education, Inc.

1.53   a. $1 + 1 \longrightarrow 2$

        b. $2 + 1 \longrightarrow 2$

1.55   a. $2Al(s) + 3Cl_2(g) \longrightarrow 2AlCl_3(s)$

        b. $2HCl(aq) + Zn(s) \longrightarrow ZnCl_2(aq) + H_2(g)$

        c. $SiO_2(s) + 2C(s) \longrightarrow SiC(s) + 2CO(g)$

Additional Problems

1.57   a. mixture   b. mixture   c. mixture

        d. pure substance   e. pure substance

1.59   d. compound   e. element

1.61   a. homogeneous   d. homogeneous

1.63

Name	Element Symbol	Group	Period	Metal or Nonmetal
Nitrogen	**N**	**5A**	**2**	**Nonmetal**
Aluminum	**Al**	3A	3	**Metal**
Sulfur	S	**6A**	3	Nonmetal
Phosphorus	P	5A	3	**Nonmetal**

1.65.   a. 1 titanium, 2 oxygens

        b. 1 nitrogen, 3 hydrogens

        c. 2 sodiums, 1 carbon, 3 oxygens

        d. 2 nitrogens, 1 oxygen

        e. 1 potassium, 1 oxygen, 1 hydrogen

1.67   a. smaller   b. larger   c. smaller   d. larger

1.69   a. 1000   b. 1   c. 1

        d. 10   e. 1,000,000

1.71   5000 strides

1.73   1 L = 1000 mL, and 2 L = 2000 mL

1.75   a.

        $\dfrac{\text{1 raisin}}{\text{1g}}$

        b. 45 raisins

1.77   a. 651,000   b. 0.00450   c. 6.67   d. 2000

1.79.   Bag 2

1.81   385 L

1.83   700 g

1.85   24 °C

1.87   a. potential   b. kinetic   c. potential   d. potential

1.89 5.4×10^5 J

1.91   0.1 cal

1.93   The copper cookware. Copper has the lowest specific heat of the three metals. All other factors being equal, it will take the copper cookware the shortest time to heat up, which will result in a shorter overall cooking time and less energy use.

Copyright © 2014 Pearson Education, Inc.

1.95   The gas particles contain more kinetic energy than the liquid particles because they are moving with higher velocities. In a gas, the particles are far apart and can move more freely.

1.97   Balance B is more accurate. Balance A is more precise.

1.99   Approximately 1.25 mL equals ¼ teaspoon in each dose.

1.101 In a solid, the particles are packed together more rigidly than in a liquid; their movement is more limited in a solid.

1.103 a. chemical reaction   b. physical change   c. chemical reaction
       d. chemical reaction   e. chemical reaction

1.105 a. $LiOH(s) + CO_2(g) \longrightarrow LiHCO_3 (s)$ (balanced)

       b. $2HCl(aq) + Mg(s) \longrightarrow MgCl_2(aq) + H_2(g)$

       c. $2Fe(s) + 3S(s) \longrightarrow Fe_2S_3(s)$

1.107 $O_2(g) + 2NO(g) \longrightarrow 2NO_2(g)$

Challenge Problems

1.109 Because the density of healthy blood is greater than the density of the copper sulfate solution, healthy blood will drop to the bottom of the tube. This happens within a few seconds.

1.111 Stepping on the brake pedal when no air is present causes the particles of the liquid to be compressed (move closer together). Because the particles in a liquid are so close together, very little pressure on the brake pedal is required to make the car stop. With air in the brake lines, depressing the pedal first compresses the gas particles, which have much more space between them. The air must be compressed with maximum pressure before the liquid is compressed to stop the car.

1.113 a. compound. It contains more than one element in its formula (H_2O).
       b. homogeneous. The sugar and water particles are evenly distributed when mixed.
       c. physical change

Copyright © 2014 Pearson Education, Inc.

Atoms and Radioactivity

2.1 Atoms and Their Components

Learning Objectives

Upon completion of this material, a student should be able to do the following:
   A. Define the following key terms:

electron	**proton**	**neutron**
nucleus	**atomic mass unit (amu)**	**subatomic particle**
electron cloud		

   B. Name the kind of subatomic particles that make up an atom. (Refer to Table 2.1.)
   C. Locate the subatomic particles in an atom.
   D. Predict the mass of an atom from the number of subatomic particles.

2.2 Atomic Number and Mass Number

Learning Objectives

Upon completion of this material, a student should be able to do the following:
   A. Define the following key terms:

mass number	**atomic number**

   B. Define atomic number.
   C. Determine the mass number for a given atom.

2.3 Isotopes and Atomic Mass

Learning Objectives

Upon completion of this material, a student should be able to do the following:
   A. Define the following key terms:

isotopes	**atomic mass**

   B. Define isotope.
   C. Distinguish mass number and atomic number.

Copyright © 2014 Pearson Education, Inc.

2.4 Radioactivity and Radioisotopes

Learning Objectives

Upon completion of this material, a student should be able to do the following:
- A. Define the following key terms:

nuclear radiation	**radioactive**
radioisotope	**radioactive decay**
alpha (α) particle	**beta (β) particle**
gamma (γ) ray	**positron**
ionizing radiation	

- B. Define radioactivity.
- C. Distinguish the forms of ionizing radiation.
- D. Differentiate the penetrating power of the forms of ionizing radiation. (Refer to Table 2.5.)

2.5 Nuclear Equations and Radioactive Decay

Learning Objectives

Upon completion of this material, a student should be able to do the following:
- A. Define the following key terms:
    nuclear decay equation
- B. Write balanced nuclear decay equations for alpha, beta, gamma, and positron emissions.

2.6 Radiation Units and Half-Lives

Learning Objectives

Upon completion of this material, a student should be able to do the following:
- A. Define the following key terms:

curie (Ci)	**becquerel (Bq)**
half-life	

- B. Perform dosing calculations using radiation activity units.
- C. Determine the remaining dose of a radioactive isotope given the half-life.

2.7 Medical Applications for Radioisotopes

Learning Objectives

Upon completion of this material, a student should be able to do the following:
- A. Define the following key term:
    tracer
- B. Contrast the use of radioisotopes for the diagnosis and treatment of disease.

Copyright © 2014 Pearson Education, Inc.

Practice Test for Chapter 2

1. A certain subatomic particle is located in the nucleus and has a relative mass of 1. This particle could have:
   A. no charge.
   B. a positive charge.
   C. a negative charge.
   D. no charge or a positive charge.

2. A subatomic particle with a large mass and a charge opposite that of an electron is represented by which of the following symbols?
   A. e^-
   B. p^+
   C. n^0
   D. n

3. What does a chemist use an amu to measure?
   A. the charge of an electron
   B. the mass of an atom
   C. radiation penetrating power
   D. a mole of atoms

4. Where is most of the mass of an atom located?
   A. electron cloud
   B. in the neutrons
   C. nucleus
   D. in the protons

5. The atomic number of an atom is equal to the number of:
   A. protons
   B. neutrons
   C. neutrons plus protons
   D. electrons plus protons

6. Consider the following symbolic notation.

$$^{27}_{13}\text{Al}$$

   This isotope has:
   A. 13 protons, 13 electrons, and 27 neutrons.
   B. 14 protons, 13 electrons, and 13 neutrons.
   C. 27 protons, 13 electrons, and 13 neutrons.
   D. 13 protons, 13 electrons, and 14 neutrons.

Copyright © 2014 Pearson Education, Inc.

7. An isotope has 29 protons, 29 electrons, and 36 neutrons. Which of the following is the correct symbolic representation?

   A. $^{36}_{29}$Cu

   B. $^{29}_{36}$Cu

   C. $^{65}_{36}$Cu

   D. $^{65}_{29}$Cu

8. Gallium (Ga) has two naturally occurring isotopes, gallium-69 and gallium-71. If the atomic mass of gallium is 69.72, this indicates that the abundance of gallium-69

   A. is greater than that of gallium-71.
   B. is less than that of gallium-71.
   C. is about the same as that of gallium-71.
   D. A conclusion cannot be reached without more data.

9. The following represents one of the element blocks on the periodic table.

   The atomic mass of this element is:
   A. 26 amu
   B. 55.85 amu
   C. 81.85 amu
   D. 56 amu

10. The following represents one of the element blocks on the periodic table.

   This element consists of two naturally occurring isotopes, chlorine-35 and chlorine-37. Which of the following correctly describes the isotopic composition of this element?

   A. The majority of any sample consists of chlorine-35.
   B. The majority of any sample consists of chlorine-37.
   C. Naturally occurring samples consist of about a 50:50 mixture of chlorine-35 and chlorine-37.

11. Which of the following types of radiation has the **least** penetrating power?
   A. α
   B. β
   C. $^{0}_{-1}$e

Copyright © 2014 Pearson Education, Inc.

D. γ

12. Which of the following types of radiation has a positive charge?

    A. α
    B. β
    C. $_{-1}^{0}e$
    D. γ

13. Which of the following types of radiation has approximately the same mass as a beta particle?

    A. α
    B. $_{0}^{0}\gamma$
    C. $_{1}^{0}e$
    D. $_{0}^{1}n$

14. Which of the following types of radiation is involved in Positron Emission Tomography (PET)?

    A. α
    B. $_{0}^{0}\gamma$
    C. $_{1}^{0}e$
    D. Both $_{0}^{0}\gamma$ and $_{1}^{0}e$

15. Which of the following most closely resembles an X-ray?

    A. α
    B. β
    C. $_{-1}^{0}e$
    D. γ

16. Which of the following types of radiation will penetrate tissue and bone?

    A. α
    B. β
    C. X-rays
    D. γ

17. Wearing a laboratory coat would protect one from which of the following types of radiation?

    A. α
    B. β
    C. $_{-1}^{0}e$
    D. γ

18. A thick sheet of lead would **not** provide protection against the effects of which of the following types of radiation?

    A. α
    B. β
    C. $_{-1}^{0}e$
    D. γ

Copyright © 2014 Pearson Education, Inc.

E. A thick sheet of lead would provide protection against the effects of all of these.

19. Which of the following equations could represent the preparation of a radioisotope in a laboratory?
    A. $^{234}_{91}\text{Pa} \longrightarrow \, ^{234}_{92}\text{U} + \, ^{0}_{-1}\text{e}$
    B. $^{58}_{26}\text{Fe} + \, ^{1}_{0}\text{n} \longrightarrow \, ^{59}_{26}\text{Fe}$
    C. $^{30}_{15}\text{P} \longrightarrow \, ^{30}_{14}\text{Si} + \, ^{0}_{1}\text{e}$
    D. $^{160}_{74}\text{W} \longrightarrow \, ^{156}_{72}\text{Hf} + \, ^{4}_{2}\text{He}$

20. An isotope containing 9 protons and 9 neutrons is used in PET scanning. Which of the following correctly characterizes this isotope?
    A. fluorine-18
    B. emits $^{0}_{1}\text{e}$
    C. mass number = 18
    D. A and C
    E. A, B, and C

21. When writing a nuclear decay equation, which of the following is correct?
    A. The sum of the mass numbers on either side of the equation must be the same.
    B. The sum of the atomic numbers on either side of the equation must be the same.
    C. The general form of the equation is: radioactive nucleus → new nucleus + radiation emission.
    D. All of these describe writing a correct nuclear decay equation.

22. Consider the following nuclear equation.
$$^{66}_{29}\text{Cu} + \, ^{0}_{-1}\text{e} \longrightarrow \text{X}$$
    In the equation, X is:
    A. $^{66}_{30}\text{Zn}$
    B. $^{66}_{28}\text{Ni}$
    C. $^{66}_{28}\text{Cu}$
    D. $^{65}_{29}\text{Cu}$

23. Consider the following nuclear equation.
$$^{212}_{84}\text{Po} + \text{X} \longrightarrow \, ^{216}_{86}\text{Rn}$$
    In the equation, X is:
    A. $^{4}_{2}\text{He}$
    B. $^{0}_{1}\text{n}$
    C. $^{0}_{1}\text{e}$
    D. $^{0}_{-1}\text{e}$

Copyright © 2014 Pearson Education, Inc.

24. Consider the following nuclear equation.

$$^{230}_{90}\text{Th} \longrightarrow {}^{230}_{90}\text{Th} + X$$

In the equation, X is:

A. ^4_2He

B. ^0_1n

C. ^0_1e

D. $^0_{-1}\text{e}$

E. $^0_0\gamma$

25. An 8.0 mL sample of a radioactive isotope used in a diagnostic procedure contains 12 mCi, how many becquerels are present in the sample?
   A. 12 Bq
   B. 1.2×10^3 Bq
   C. 1.2×10^3 Bq
   D. 4.4×10^8 Bq
   E. 4.4×10^{14} Bq

26. An 8.0 mL sample of a radioactive isotope used in a diagnostic procedure contains 12 mCi. If the patient is to receive 2.8 mCi, what volume in milliliters should be injected?
   A. 8.0 mL
   B. 1.9 mL
   C. 0.67 mL
   D. 4.2 mL
   E. 12 mL

27. Chromium-51 has a half-life of 28 days. If a 222 g sample is received in a clinical lab, how long will it take until only 27.8 g of radioactive chromium-51 remains?
   A. 28 days
   B. 3 days
   C. 56 days
   D. 84 days

28. Technicium-99m is a useful imaging tool for medical diagnosis. This isotope has a short half-life of 6 hours. If a patient receives a dose of 28 mCi, how much radioactivity will remain when the patient goes home 24 hours later?
   A. 4.0 mCi
   B. 7.0 mCi
   C. 3.5 mCi
   D. 1.8 mCi

29. A patient receives a 50 µCi does of a gamma emitter. Which of the following is most likely correct?
   A. The gamma emitter is a tracer.
   B. The procedure is diagnostic in nature.

Copyright © 2014 Pearson Education, Inc.

C. The procedure is therapeutic in nature.
D. A and B
E. A and C

30. Consider the following data.

Radioisotope	Half-Life
Hydrogen-3	12.3 years
Iron-59	46 days
Iodine-131	8 days
Iodine-123	13.3 hours

Which isotope would be the **least** likely to be used in a medical application?
A. hydrogen-3
B. iron-59
C. iodine-131
D. iodine-123
E. All would be likely candidates.

Copyright © 2014 Pearson Education, Inc.

Answers

1. D 2. B 3. B 4. C 5. A 6. D 7. D 8. A 9. B 10. A
11. A 12. A 13. C 14. D 15. D 16. D 17. A 18. D 19. B 20. E
21. D 22. B 23. A 24. E 25. D 26. B 27. D 28. D 29. D 30. A

Copyright © 2014 Pearson Education, Inc.

Chapter 2 – Solutions to Odd-Numbered Problems

Practice Problems

2.1   Protons and neutrons are located in the nucleus (center) of an atom, and the electrons are found in a cloud outside of the nucleus called the electron cloud.

2.3   The mass of an electron is about 2000 times smaller than that of a proton.

2.5   a. The number of protons is the atomic number.
       b. The number of neutrons is the mass number minus the atomic number.
       c. The number of electrons is the same as the number of protons in an atom.

2.7   a. oxygen, O   b. magnesium, Mg   c. neon, Ne   d. copper, Cu
       e. silver, Ag

2.9   14

2.11   a. 35 protons, 45 neutrons, 35 electrons
       b. 11 protons, 12 neutrons, 11 electrons

2.13   a. $_2^4 He$ b. $_{17}^{35} Cl$ c. $_{16}^{32} S$ d. $_{55}^{133} Cs$

2.15   a. 8 protons, 10 neutrons, 8 electrons
       b. 20 protons, 20 neutrons, 20 electrons
       c. 47 protons, 61 neutrons, 47 electrons
       d. 82 protons, 125 neutrons, 82 electrons

2.17   The isotope carbon-12 contains exactly 6 protons and 6 neutrons. The atomic mass as seen on the periodic table is an average mass taking into consideration the abundance of all the carbon isotopes. Because about 1% of the carbon isotopes are carbon-13, the atomic mass is slightly higher than 12 amu.

2.19   a. Magnesium-24 has 12 neutrons, magnesium-25 has 13 neutrons, magnesium-26 has 14 neutrons.
       b. $_{12}^{24} Mg$, $_{12}^{25} Mg$, $_{12}^{26} Mg$
       c. magnesium-24

2.21   An alpha particle is a helium nucleus. There are no electrons in an alpha particle.

2.23   gamma

2.25   a. $_6^{14} C \rightarrow _7^{14} N + _{-1}^0 e$

       b. $_{84}^{212} Po \rightarrow _{82}^{208} Pb + _2^4 He$

       c. $_{29}^{66} Cu \rightarrow _{30}^{66} Zn + _{-1}^0 e$

       d. $_6^{11} C \rightarrow _5^{11} B + _1^0 e$

2.27   The half-lives are shorter, and they are prepared in the lab.

2.29

$$50 \; \cancel{mCi} \; \times \frac{20 \, mL}{250 \; \cancel{mCi}} = 4 \, mL$$

2.31

$$400 \; \mu Ci \; \underset{\text{1 half-life}}{\overset{\text{8 days}}{\rightarrow}} \; 200 \; \mu Ci \; \underset{\text{2 half-lives}}{\overset{\text{16 days}}{\rightarrow}} \; 100 \; \mu Ci \; \underset{\text{3 half-lives}}{\overset{\text{24 days}}{\rightarrow}} \; 50 \; \mu Ci \; \underset{\text{4 half-lives}}{\overset{\text{32 days}}{\rightarrow}} \; 25.0 \; \mu Ci$$

2.33   Ca-47 will concentrate in bone.

2.35

$$40 \; \mu Ci \; \xrightarrow[\text{1 half-life}]{} 20 \; \mu Ci \; \xrightarrow[\text{2 half-lives}]{} 10 \; \mu Ci \; \xrightarrow[\text{3 half-lives}]{} 5.0 \; \mu Ci$$

Copyright © 2014 Pearson Education, Inc.

After 3 half-lives (84 days), the patient will have less than one-tenth the original dose.

Additional Problems

2.37 a. neutron   b. protons   c. isotopes

2.39 a. 79 protons, 79 electrons
      b. 30 protons, 30 electrons
      c. 29 protons, 29 electrons

2.41 a. 26 protons, 29 neutrons, 26 electrons
      b. 7 protons, 8 neutrons, 7 electrons
      c. 24 protons, 28 neutrons, 24 electrons
      d. 56 protons, 81 neutrons, 56 electrons

2.43

Symbol	Number of Protons	Number of Neutrons	Number of Electrons	Mass Number	Name
$^{1}_{1}H$	1	0	1	1	Hydrogen-1
$^{24}_{12}Mg$	12	12	12	24	Magnesium-12
$^{9}_{4}Be$	4	5	4	9	Beryllium-9

2.45 a. alpha particle   b. beta particle   c. gamma radiation

2.47 a. $^{32}_{15}P$ b. $^{60}_{27}Co$ c. $^{51}_{24}Cr$

2.49 (answers in boldface)

Isotope Name	Symbolic Notation	Number of Protons	Number of Neutrons	Mass Number	Medical Use
Thallium-201	$^{201}_{81}Tl$	**81**	**120**	**201**	Tumor Imaging
Iodine-123	$^{123}_{53}I$	53	**70**	**123**	Thyroid Imaging
Xenon-133	$^{133}_{54}Xe$	54	**79**	133	Pulmonary Ventilation Imaging
Fluorine-18	$^{18}_{9}F$	**9**	**9**	**18**	Positron Emission Tomography (PET) Imaging

2.51 a. $^{15}_{8}O \rightarrow {}^{15}_{7}N + {}^{0}_{1}e$

      b. $^{46}_{23}V \rightarrow {}^{46}_{24}Cr + {}^{0}_{-1}e$

      c. $^{234}_{92}U \rightarrow {}^{230}_{90}Th + {}^{4}_{2}He$

      d. $^{8}_{4}Be \rightarrow {}^{4}_{2}He + {}^{4}_{2}He$

2.53 a. $^{66}_{29}Cu \rightarrow {}^{66}_{30}Zn + {}^{0}_{-1}e$

      b. $^{192}_{78}Pt \rightarrow {}^{188}_{76}Os + {}^{4}_{2}He$

Copyright © 2014 Pearson Education, Inc.

c. $^{126}_{50}\text{Sn} \rightarrow ^{126}_{51}\text{Sb} + ^{0}_{-1}\text{e}$

d. $^{72}_{31}\text{Ga} \rightarrow ^{72}_{32}\text{Ge} + ^{0}_{-1}\text{e}$

2.55 $^{10}_{5}\text{B} + ^{4}_{2}\text{He} \rightarrow ^{12}_{7}\text{N} + 2\,^{1}_{0}\text{n}$

2.57   1.3 mL

2.59   12.5 mg

2.61   13.2 hours

2.63   A cold spot indicates a diseased area of an organ; a hot spot indicates an area of the organ where the cells are dividing rapidly or a cancerous growth.

Challenge Problems

2.65   1.56%

2.67   Because a person with a more active brain uses more glucose, more FDG will be present in the brain. The PET image would be brighter in more areas than for the normal person.

Copyright © 2014 Pearson Education, Inc.

Compounds—
Putting Particles Together

3.1   Electron Arrangements and the Octet Rule

Learning Objectives

Upon completion of this material, a student should be able to do the following:
   A. Define the following key terms:

valance shell	**noble gases**
valence electrons	**octet rule**

   B. Predict the number of valence electrons and energy levels for the main-group elements in the first four rows.
   C. Recognize the unique stability associated with a valence shell containing 8 electrons.

3.2   In Search of an Octet, Part 1: Ion Formation

Learning Objectives

Upon completion of this material, a student should be able to do the following:
   A. Define the following key terms:

ions	**isoelectronic**
anion	**polyatomic ion**
cation	

   B. Predict the ionic charge of a main-group element using the periodic table.
   C. Distinguish the name of an ion from its corresponding atom name.
   D. Gain familiarity with polyatomic ions and their charges.

3.3   Ionic Compounds – Electron Give and Take

Learning Objectives

Upon completion of this material, a student should be able to do the following:
   A. Define the following key terms:

ionic bond	**ionic compound**

   B. Predict the ionic charges present in an ionic compound.
   C. Predict the ionic charge of a transition metal using the compound's formula and the anionic charge.
   D. Name ionic compounds given the formula.
   E. Write the formula for ionic compounds given the name.

Copyright © 2014 Pearson Education, Inc.

3.4 In Search of an Octet, Part 2: Covalent Bonding

Learning Objectives

Upon completion of this material, a student should be able to do the following:

A. Define the following key terms:

covalent bond	double bond
covalent compound	triple bond
molecule	molecular formula
bonding pair	Lewis structure
lone pair	binary compound
single bond	

B. Distinguish ionic and covalent compounds.

C. Establish the relationship between the number of valence electrons present in the Period 1-3 nonmetals and Group 7A elements and the number of bonds that the atom typically makes in a molecule.

D. Draw Lewis structures for covalent compounds containing C, O, N, H, and the halogens (Group 7A).

E. Name binary covalent compounds given the formula.

F. Write the formula for a binary covalent compound given the name.

3.5 The Mole: Counting Atoms and Compounds

Learning Objectives

Upon completion of this material, a student should be able to do the following:

A. Define the following key terms:

mole	equivalent units
molar mass	dalton
Avogadro's number (N)	

B. Describe the mole unit and Avogadro's number.

C. Calculate molar mass for a compound.

D. Convert among the units of mole, number of particles, and gram.

3.6 Getting Covalent Compounds into Shape

Learning Objectives

Upon completion of this material, a student should be able to do the following:

A. Define the following key terms:

tetrahedral	valence-shell electron-pair repulsion (VSEPR)
nonbonded pairs	charge clouds

B. Predict the molecular shapes of small molecules using VSEPR.

C. Determine the effect of lone pair electrons on molecular shape.

Copyright © 2014 Pearson Education, Inc.

3.7 Electronegativity and Molecular Polarity

Learning Objectives

Upon completion of this material, a student should be able to do the following:

    A. Define the following key terms:

          electronegativity   **polar molecule**

          polar covalent bond   **nonpolar molecule**

          nonpolar covalent bond

    B. Predict covalent bond polarity based on electronegativity.

    C. Predict molecular polarity from predicted bond polarities.

Copyright © 2014 Pearson Education, Inc.

Practice Test for Chapter 3

1. Which of the following represents the number of electrons in each energy level of magnesium?

   A.

N	Number of Electrons
1	8
2	2
3	

   B.

N	Number of Electrons
1	2
2	8
3	

   C.

N	Number of Electrons
1	2
2	8
3	2

   D.

N	Number of Electrons
1	2
2	8
3	12

2. How many electrons constitute a full energy level if $n = 2$?
   A. 2
   B. 6
   C. 8
   D. 10

3. How many valence electrons are there in an atom of phosphorus (P)?
   A. 3
   B. 5
   C. 15
   D. 16

4. An ion containing 38 protons and 36 electrons could be represented and named as:
   A. Sr^{2+}, strontium ion.
   B. Kr^{2-}, kryptonide.
   C. Sr^{2-}, strontide ion.
   D. Kr^{2+}, krypton ion.

Copyright © 2014 Pearson Education, Inc.

5. Determine the correct formula of the ionic compound formed from the following combination of potassium ions and nitride ions.

   A. KN
   B. K_2N
   C. KN_3
   D. K_3N

6. The formula for barium carbonate is:
   A. BaC.
   B. $BaCO_3$.
   C. $Ba(HCO_3)_2$.
   D. Ba_2C.

7. Which of the following are, respectively, the correct name and composition of the ionic compound formed between ions of nickel(III) and sulfide?
   A. nickel(II) sulfate, 1 Ni^{2+} and S^{2-}
   B. nickel(III) sulfate, 3 Ni^{3+} and 2 S^{2-}
   C. nickel(III) sulfide, 2 Ni^{3+} and 3 S^{2-}
   D. nickel(II) sulfate, 1 Ni^{2+} and S^{2-}

8. The correct Lewis structure for OF_2 would contain:

   A. 1 O—F bond and 1 F—F bond.
   B. 1 O=F bond and 1 F—F bond.
   C. 2 O—F bonds.
   D. 1 O—F bond and 1 F=O bond.

9. If Cl atoms replaced the Br atoms in the molecule shown, which of the following would describe the new Lewis structure?

$$:\ddot{Br}—\ddot{P}—\ddot{Br}:$$
$$:\ddot{Br}:$$

   A. There would be the same number of lone pairs.
   B. One of the bonds between P and Br would be double.
   C. P would no longer have a lone pair.
   D. The bonds would become ionic.
   E. All of the above.

Copyright © 2014 Pearson Education, Inc.

10. Which of the following pairs contains an ionic compound and a covalent compound?
    A. NH_3 and NH_4Cl
    B. PCl_3 and $ClBr$
    C. AlI_3 and SF_2
    D. both A and C

11. Based on Lewis structures, which of the following compounds is **not** likely to exist?
    A. CCl_4
    B. H_3S
    C. OCl_2
    D. PI_3

12. The correct name for P_2O_5 would be:
    A. diphosphorous pentoxide.
    B. phosphorous oxide.
    C. phosphorous(II) oxide.
    D. pentaphosphorous dioxide.

13. The formula for dinitrogen monoxide is:
    A. NO.
    B. N_2O.
    C. NO_2.
    D. N_2O_3.

14. What is the mass of 3.50 moles of sodium?
    A. 80.5 g
    B. 3.50 g
    C. 22.99 g
    D. 6.57 g

15. How many moles of platinum atoms are in 10.0 g of Pt?
    A. 195 mol of Pt
    B. 0.0513 mol of Pt
    C. 3.08×10^{22} mol of Pt
    D. 1950 mol of Pt

16. How many calcium atoms are in a beaker containing 22.5 g of Ca?
    A. 0.561 atoms
    B. 5.42×10^{26} atoms
    C. 1.07×10^{24} atoms
    D. 3.38×10^{23} atoms

Copyright © 2014 Pearson Education, Inc.

17. Which of the following contains the largest number of atoms?
    A. 1.00 mol of K
    B. 1.00 mol of Li
    C. 1.00 mol of Na
    D. All contain the same number of atoms.

18. Which of the following contains the largest number of atoms?
    A. 5.00 g of copper
    B. 5.00 g of carbon
    C. 5.00 g of cesium
    D. All contain the same number of atoms.

19. What is the VSEPR form of the following molecule?

$$SO_3$$

    A. AB_2
    B. AB_3
    C. AB_2N_2
    D. AB_3N

20. What is the shape of a molecule of NBr_3?
    A. tetrahedral
    B. pyramidal
    C. trigonal planar
    D. bent

21. In which of the following molecules is the shape around each carbon atom tetrahedral?
    A.

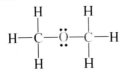

    B.

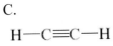

    C.

H—C≡C—H

    D.

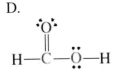

Copyright © 2014 Pearson Education, Inc.

22. In which of the following bonds will the Br atom exist with a δ–?
    A. Br—O
    B. Br—F
    C. Br—N
    D. Br—Te

23. A student wrote the following as the Lewis structure for C_2H_4.

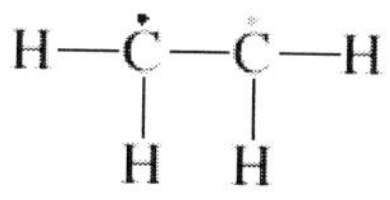

H—Ċ—Ċ—H
   | |
   H   H

   What error(s), if any, was(were) made?
   A. The number of atoms of C and H are incorrect.
   B. The H atoms do not form the correct number of bonds.
   C. The C atoms do not have an octet of electrons.
   D. All of the above are errors present in the structure.
   E. The structure is correct as written.

24. In which of the following bonds will the dipole moment arrow (⟶) point toward N?
    A. N—S
    B. N—O
    C. N—Cl
    D. N—F

25. Which of the following molecules contains polar bonds?
    A. CI_4
    B. NCl_3
    C. OF_2
    D. All contain polar bonds.

26. The most polar bond among those listed would be:
    A. F—F.
    B. O—F.
    C. C—F.
    D. N—F.

Copyright © 2014 Pearson Education, Inc.

27. Of the compounds listed, which would consist of nonpolar molecules?

A.

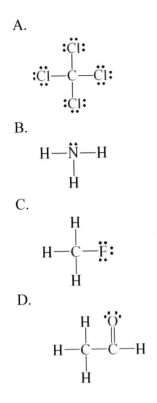

B.

H—N̈—H
|
H

C.

```
    H
    |
H—C—F̈:
    |
    H
```

D.

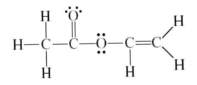

28. In the following molecule shown here, how many of the carbon atoms have trigonal planar shape?

A. 1
B. 2
C. 3
D. 4
E. none

Copyright © 2014 Pearson Education, Inc.

29. For the following molecule, the arrow for the molecular dipole should be drawn as:

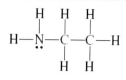

A.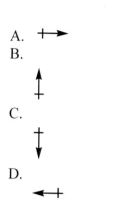

B.

C.

D.

E. The molecule does not have a dipole.

30. In the following structure, the shape around the N atom is:

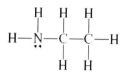

A. tetrahedral.
B. pyramidal.
C. trigonal planar.
D. linear.

Copyright © 2014 Pearson Education, Inc.

Answers

1. C   2. C   3. B   4. A   5. D   6. B   7. C   8. C   9. A   10. D
11. B   12. A   13. B   14. A   15. B   16. D   17. D   18. B   19. B   20. B
21. A   22. D   23. C   24. A   25. D   26. C   27. A   28. C   29. D   30. B

Copyright © 2014 Pearson Education, Inc.

Chapter 3 – Solutions to Odd-Numbered Problems

Practice Problems

3.1 2 electrons, 8 electrons

3.3 a. 2 e^- in first energy level
   b. 2 e^- in first energy level, 4 e^- in second energy level
   c. 2 e^- in first energy level, 8 e^- in second energy level, 1 e^- in third energy level
   d. 2 e^- in first energy level, 8 e^- in second energy level

3.5 a. 6   b. 4   c. 5   d. 1

3.7 In atoms, the number of protons and electrons are the same and there is no net charge. In a cation, there are more protons than electrons, giving the cation a positive charge.

3.9 In naming an anion, the last several letters of the nonmetal element name are dropped and the suffix ide is applied.

3.11 a. 20 p, 18 e^- b. 53 p, 54 e^- c. 16 p, 18 e^- d. 30 p, 28 e^-

3.13 a. calcium ion   b. iodide   c. sulfide   d. zinc ion

3.15 a. fluoride, F^- b. chromium(III) ion, Cr^{3+}

3.17 a. ammonium ion   b. acetate   c. cyanide

3.19 a. $AuCl^3$, gold(III) chloride
   b. $CaCO^3$, calcium carbonate
   c. $Mg(OH)^2$, magnesium hydroxide

3.21 a. one Au^{3+}, three Cl^- b. one Ca^{2+}, one CO32–   c. one Mg^{2+}, two OH–

3.23 a. sodium carbonate, $Na2CO^3$
   b. iron(II) carbonate, $FeCO^3$
   c. aluminum carbonate, $Al^2(CO^3)^3$

3.25 a. two Na^+, one CO_3^{2-}
   b. one Fe^{2+}, one CO_3^{2-}
   c. two Al^{3+}, three CO_3^{2-}

3.27

(a)   (b)

(c)   (d)

3.29 a. Carbon prefers four bonds, hydrogen only one each, so CH_2 is unlikely to exist.
   b. Nitrogen prefers three bonds, hydrogen only one each, so NH_2 is unlikely to exist.
   c. These halogens could form one covalent bond like Cl_2, so BrCl could exist.
   d. Sulfur will form two bonds, hydrogen one each, so H_2S could exist.

3.31 a. covalent   b. ionic   c. ionic   d. covalent
   e. covalent   f. ionic   g. covalent   h. ionic
   i. covalent   j. covalent

3.33 SO_3, diphosphorus pentoxide, SeF_4, carbon monoxide, N_2O_3

3.35 a. They are equal; 6.02 x 1023 atoms.
   b. Gold weighs more, 197.0 g of gold versus 107.9 g of silver.

Copyright © 2014 Pearson Education, Inc.

3.37   a. 1.51 x 1023 atoms b. 0.019 mole   c. 2.57 x 1023 atoms

3.39   a. 120.38 g/mole   b. 180.16 g/mole   c. 44.01 g/mole

3.41   1.48 x 1021 molecules

3.43   a.
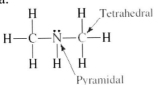

b.
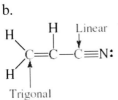

c.
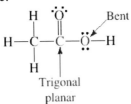

d.
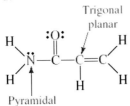

3.45

    a.

$$H-Cl$$
$$\longmapsto$$
Polar

Because this molecule contains one
bond, the bond and molecular dipoles
are the same.

    b.

nonpolar

$$H-\overset{\overset{\displaystyle H}{|}}{\underset{\underset{\displaystyle H}{|}}{C}}-H$$

Nonpolar
C–H bonds are considered nonpolar.

Copyright © 2014 Pearson Education, Inc.

c.

$$:\ddot{C}l\!-\!\ddot{N}\!-\!\ddot{C}l: \quad \uparrow$$
$$\underset{:\ddot{C}l:}{|}$$

Polar

N–Cl bonds are nonpolar (same electronegativity), but
the pair of electrons on the N gives the molecule polarity.

d.

$$\ddot{S}\!=\!C\!=\!\ddot{O}$$
$$\xrightarrow{\quad}$$

Polar

C–O bonds are polar and C–S bonds are not. The
molecular dipole is shown in orange.

e.

Polar

S–H bonds are polar. This molecule has a bent shape.
The lone pairs of electrons are on one side of the
molecule in three dimensions. The molecular dipole is
shown in orange.

f.

Polar

C–Br bonds are polar. Considering the tetrahedral shape,
the bromine side of the molecule is the negative side.

Additional Problems

3.47   a. two electrons, Group 2A
        b. seven electrons, Group 7A
        c. six electrons, Group 6A
        d. five electrons, Group 5A
        e. two electrons, Group 2A
        f. seven electrons, Group 7A

3.49	a. 4	b. 7	c. 1	d. 3
3.51	a. 4	b. 1	c. 1	d. 3
3.53	a. fewer	b. valence	c. positive	
3.55	a. 17 p, 18 e^-	b. 26 p, 23 e^-	c. 24 p, 18 e^-	d. 7 p, 10 e^-
	e. 11 p, 10 e^-	f. 1 p, 0 e^-		
3.57	b. iron(III)	c. chromium(VI)	d. nitride	f. hydrogen ion, also proton
3.59	a. lithium ion, Li+	b. bromide, Br–		
3.61	a. Kr	b. Ne	c. Ar	d. Ar

Copyright © 2014 Pearson Education, Inc.

3.63   a. $C_2H_3O_2^-$ b. HCO_3^- c. nitrate   d. cyanide

3.65   a. $(NH_4)_2S$ b. NH_4Cl c. $(NH_4)_2SO_4$ d. NH_4OH

3.67   a. NaOH   b. Al_2O_3 c. KNO_3

3.69   a. sodium oxide   b. barium sulfate   c. copper(II) chloride

      d. magnesium nitrate   e. iron(III) oxide   f. potassium fluoride

3.71 I^-

3.73 $Na_2S_2O_5$

3.75   a. 4   b. 1   c. 3   d. 2

3.77   a. covalent, carbon tetrabromide

      b. covalent, silicon dioxide

      c. ionic, magnesium bromide

      d. covalent, nitrogen trichloride

      e. ionic, chromium(III) chloride

3.79   Ionic formulas are written for ionic compounds and use the smallest ratio of ions; molecular formulas are written for covalent compounds and indicate the exact number of atoms in the molecule.

3.81   a.   H—C≡N:

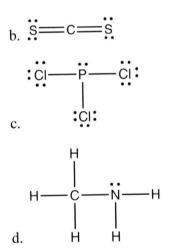

3.83   a. selenium dioxide

      b. silicon tetrafluoride

      c. tetraphosphorus trisulfide

      d. oxygen difluoride

3.85   A covalent bond results when two atoms share one or more pairs of electrons. An ionic bond is the attraction between two or more ions, which are formed by the loss and gain of electrons.

3.87   a. 4.03 g   b. 27.8 g   c. 404 g   d. 184 g

3.89   3 x 1017 atoms of carbon

Copyright © 2014 Pearson Education, Inc.

3.91

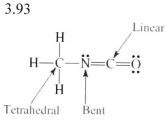

Trigonal planar, 120°

Pyramidal, <109.5°

Bent, <109.5°

Tetrahedral, 109.5°

Bent, <109.5°

3.93

Linear

Tetrahedral   Bent

3.95   a. S   b. N   c. Cl   d. N

3.97

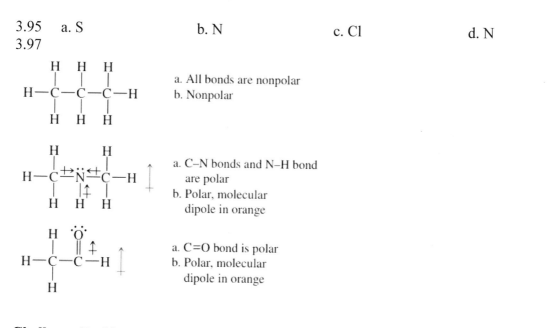

a. All bonds are nonpolar
b. Nonpolar

a. C–N bonds and N–H bond are polar
b. Polar, molecular dipole in orange

a. C=O bond is polar
b. Polar, molecular dipole in orange

Challenge Problems

3.99 $Ca_{10}(PO_4)_6(OH)_2$
3.101 9.60×10^{22} atoms of Pt

Copyright © 2014 Pearson Education, Inc.

3.103 a.

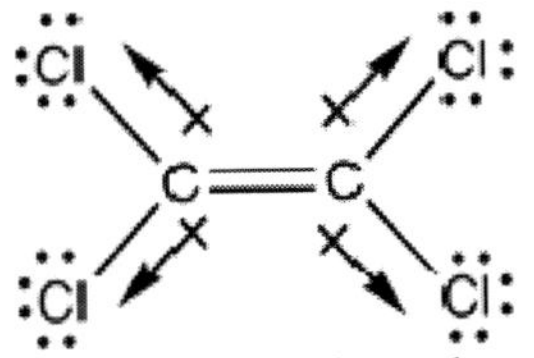

b. Both carbons are trigonal planar and the bond angle is 120°.

c. C—Cl bonds are polar, dipoles shown at left.

d. The polar bonds are pulling outward equally and oppositely, so the molecule has no molecular dipole and is nonpolar.

Copyright © 2014 Pearson Education, Inc.

Introduction to Organic Compounds

4.1 Alkanes: The Simplest Organic Compounds

Learning Objectives

Upon completion of this material, a student should be able to do the following:

A. Define the following key terms:

organic compounds   alkanes
biomolecules   saturated hydrocarbons
organic chemistry   straight-chain alkane
inorganic compounds   cycloalkane

B. Distinguish organic and inorganic compounds.

C. Define the terms saturated and unsaturated hydrocarbon.

D. Compare the molecular formulas for straight-chain alkanes and cycloalkanes.

4.2 Representing the Structures of Organic Compounds

Learning Objectives

Upon completion of this material, a student should be able to do the following:

A. Define the following key terms:

condensed structural formula   skeletal structure

B. Draw organic compounds as Lewis, condensed, and skeletal structures.

4.3 Families of Organic Compounds—Functional Groups

Learning Objectives

Upon completion of this material, a student should be able to do the following:

A. Define the following key terms:

heteroatoms   aromatic compounds
functional group   aromaticity
carbonyl   resonance hybrid
carboxylic acid   monounsaturated
alkene   polyunsaturated
terpenes   fatty acids
unsaturated hydrocarbons   saturated fatty acids
alkyne

B. Identify common functional groups in organic molecules.

C. Characterize the unsaturated hydrocarbons alkenes, alkynes, and aromatics.

D. Draw saturated fatty acids in skeletal structure.

Copyright © 2014 Pearson Education, Inc.

4.4   Nomenclature of Simple Alkanes

Learning Objectives

Upon completion of this material, a student should be able to do the following:
- A. Define the following key terms:

  branched-chain alkanes   **alkyl groups**

  substituents   **haloalkanes**
- B. Name branched-chain alkanes, haloalkanes, and cycloalkanes using IUPAC naming rules.

4.5   Isomerism in Organic Compounds

Learning Objectives

Upon completion of this material, a student should be able to do the following:
- A. Define the following key terms:

  isomer   **unsaturated fatty acids**

  structural isomers   **omega number**

  conformational isomer   **essential fatty acids**

  conformers   **enantiomer**

  stereoisomer   **chiral**

  cis–trans stereoisomers   **chiral center**

  restricted rotation

- B. Distinguish structural isomers from conformational isomers.
- C. Identify cis and trans isomers in cycloalkanes and alkenes.
- D. Draw unsaturated fatty acids in skeletal structure.
- E. Locate chiral centers in organic molecules.

Copyright © 2014 Pearson Education, Inc.

Practice Test for Chapter 4

1. Which of the following compounds would be classified as inorganic?

   A. CH_3COOH
   B. CH_2Cl_2
   C. $NaHCO_3$
   D. C_6H_6

2. What is the molecular formula for the following structure?

   A. C_6H_{10}
   B. C_6H_{15}
   C. C_6H_8
   D. C_6H_{12}

3. Which of the choices does **not** represent the same molecule as shown here?

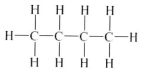

   A.

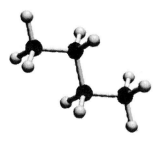

   B.

C.

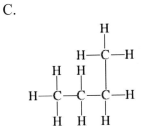

D.

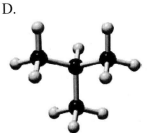

4. Which of the terms does **not** describe the following structure?

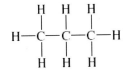

   A. hydrocarbon
   B. alkane
   C. nonpolar
   D. unsaturated

5. Which of the given formulas represents a skeletal structure?
   A. C_4H_{12}
   B.

   C.

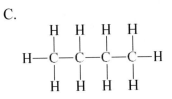

   D. $CH_3CH_2CH_2CH_3$

Copyright © 2014 Pearson Education, Inc.

6. Pentane could be represented as:

   A.

   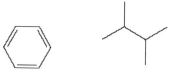

   B. $CH_3CH_2CH_2CH_2CH_3$
   C.

   D. C_6H_{14}

7. Identify the family of hydrocarbons present in each of the following, respectively.

   H—C≡C—H

   A. aromatic, alkane, alkene
   B. alkene, alkane, alkyne
   C. aromatic, alkane, alkyne
   D. cycloalkene, alkane, alkyne

8. The fundamental feature that distinguishes an alkane from an alkene from an alkyne is
   A. the type of carbon to carbon bonds.
   B. the presence of heteroatoms.
   C. the presence of rings.
   D. aromaticity.

9. Which of the following functional groups contains a heteroatom?
   A. alkyne
   B. aldehyde
   C. aromatic
   D. alkene

10. Identify, from left to right, the functional groups in the following molecule that contain a heteroatom.

    A. alcohol, amine, carboxylic acid
    B. alcohol, amine, carboxylate
    C. phenol, amine, carboxylic acid
    D. phenol, amide, carboxylic acid

Copyright © 2014 Pearson Education, Inc.

11. The members of the following pair are:

    A. not related.
    B. structural isomers.
    C. conformational isomers.
    D. stereoisomers.

12. The members of the following pair are:

$$CH_3CH_2CH_2CHCH_3$$
$$|$$
$$CH_3$$

$$CH_3CH_2CH_2CH_2CH_2CH_3$$

    A. not related.
    B. structural isomers.
    C. conformational isomers.
    D. stereoisomers.

13. The members of the following pair are:

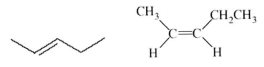

    A. not related.
    B. structural isomers.
    C. conformational isomers.
    D. stereoisomers.

14. The members of the following pair are:

$$CH_3CH_2$$
$$|$$
$$CH_2CH_2CH_2$$
$$|$$
$$CH_3$$

$$CH_3CH_2CH_2CH_2CH_2CH_3$$

    A. not related.
    B. structural isomers.
    C. conformational isomers.
    D. stereoisomers.

15. A propyl alkyl group can be represented as:
    A. $CH_3CH_2—$.
    B. $CH_3\ CH_2CH_2—$.
    C. $CH_3\ CH_2CH_2CH_2—$.
    D. $CH_3CH_2CH_2CH_2CH_2—$.

Copyright © 2014 Pearson Education, Inc.

16. If the terms *iodo*, *dimethyl*, and *ene* are found in the name for a compound, which of the following most likely characterizes this compound?
    A. It is unsaturated, contains a hetero atom and a CH_3- substituent.
    B. It is saturated, contains a hetero atom and two CH_3- substituents.
    C. It is unsaturated, contains two hetero atoms and a CH_3- substituent.
    D. It is unsaturated, contains a hetero atom and two CH_3- substituents.

17. The IUPAC name for the following compound is:

$$CH_3CH_2$$
$$|$$
$$CH_2CH_2CH_2CH_3$$

    A. 1-ethylbutane.
    B. hexane.
    C. 1-butylethane.
    D. *trans*- 1-ethylbutane.

18. The IUPAC name for the following compound is:

$$CH_3$$
$$|$$
$$CH_3CH_2CCH_2CH_3$$
$$|$$
$$CH_2CH_3$$

    A. 2,2-diethylbutane.
    B. 3,3-diethylbutane.
    C. 3-ethyl-3-methylpentane.
    D. 3-methyl-3-ethylpentane.

19. The IUPAC name for the following compound is:

    A. 2-fluoro-3-methylpentane.
    B. 2-fluoro-3-ethylbutane.
    C. 2-fluoroethylbutane.
    D. 4-fluoro-3-methylpentane.

20. The IUPAC name for the following compound is:

$$CH_3CH_2CHCH=CH_2$$
$$|$$
$$CH_3$$

    A. 3-methylpentane.
    B. 3-methylpentene.
    C. 3-methyl-1-pentene.

Copyright © 2014 Pearson Education, Inc.

D. 3-methyl-4-pentene.

21. The IUPAC name for the following compound is:

A. *trans*-1-bromo-2-methylpentane.
B. *cis*-1-bromo-2-ethylcyclopentane.
C. *cis*-1-bromo-2-methylcyclopentane.
D. *trans*-1-bromo-2-ethylcyclopentane.

22. A compound with the name 2-chloro-4-methylpentane could be represented as:
A. $CH_3CHClCH_2CH(CH_3)CH_3$.
B.

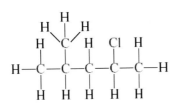

C.

D. B and C
E. A, B, and C

23. Consider the following structure.

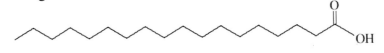

This represents:
A. a fatty acid.
B. stearic acid [18:0].
C. a lipid.
D. both A and B
E. A, B, and C

24. The following compound shown here would be classified as:

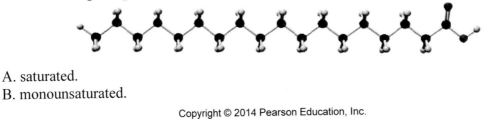

A. saturated.
B. monounsaturated.

Copyright © 2014 Pearson Education, Inc.

C. polyunsaturated.
D. essential.

25. The correct carbon designation for the following fatty acid is:

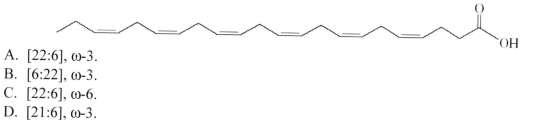

A. [22:6], ω-3.
B. [6:22], ω-3.
C. [22:6], ω-6.
D. [21:6], ω-3.

26. Which of the following pairs represents a set of enantiomers?
   A.

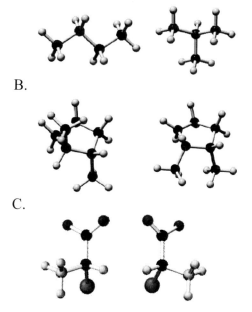

   B.

   C.

   D. B and C

27. In which type of isomer are the molecular formulas for the isomers different?
   A. conformational
   B. structural
   C. stereoisomers
   D. none of the above

Copyright © 2014 Pearson Education, Inc.

28. How many chiral carbon atoms are present in the following molecule?

A. 1
B. 2
C. 3
D. 4
E. 5

29. Which of the following molecules contains **more than one** chiral carbon atom?

A.

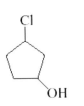

B.

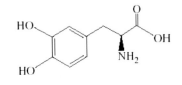

C.

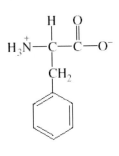

D.

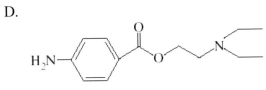

Copyright © 2014 Pearson Education, Inc.

30. Which of the following does **not** describe a pair of enantiomers?
    A. biologically equivalent
    B. mirror images
    C. contain chiral carbon atoms
    D. "handed"

Copyright © 2014 Pearson Education, Inc.

Answers

1. C   2. D   3. D   4. D   5. B   6. B   7. C   8. A   9. B   10. C

11. A   12. B   13. D   14. C   15. B   16. D   17. B   18. C   19. A   20. C

21. B   22. E   23. E   24. A   25. A   26. C   27. D   28. A   29. A   30. A

Copyright © 2014 Pearson Education, Inc.

Chapter 4 – Solutions to Odd-Numbered Problems

Practice Problems

4.1   a. cyclobutane   b. pentane   c. heptane

4.3   a. C_4H_{10} b. C_5H_{10} c. C_8H_{18}

4.5   A Lewis structure shows all atoms, bonds, and nonbonding electrons. A condensed structure shows all atoms, but as few bonds as possible.

4.7 CH_4

CH_3CH_3

$CH_3CH_2CH_3$

$CH_3CH_2CH_2CH_3$

$CH_3CH_2CH_2CH_2CH_3$

$CH_3CH_2CH_2CH_2CH_2CH_3$

$CH_3CH_2CH_2CH_2CH_2CH_2CH_3$

$CH_3CH_2CH_2CH_2CH_2CH_2CH_2CH_3$

$CH_3CH_2CH_2CH_2CH_2CH_2CH_2CH_2CH_3$

$CH_3CH_2CH_2CH_2CH_2CH_2CH_2CH_2CH_2CH_3$

4.9   Skeletal structures show bonds between carbon atoms. Since methane has only one carbon, it is not possible to draw its skeletal structure.

4.11

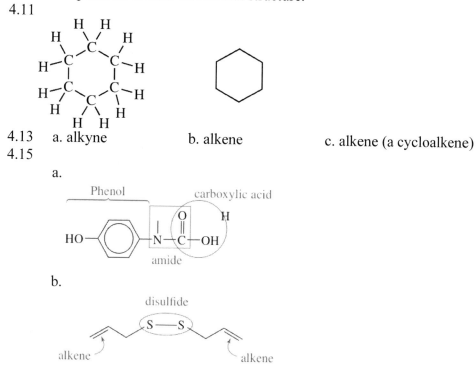

4.13   a. alkyne   b. alkene   c. alkene (a cycloalkene)

4.15

a.

b.

4.17

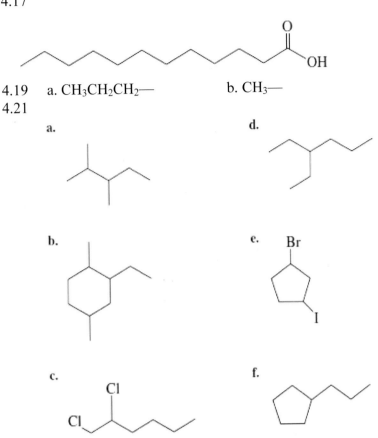

4.19   a. CH₃CH₂CH₂—   b. CH₃—
4.21

a.   d.

b.   e.   Br

c.   f.
        Cl

4.23   a. Conformational isomers have the same molecular formula and connectivity.
        b. Conformational isomers differ by rotation about one or more single bonds.
4.25   a. structural isomers   b. not related
        c. not related   d. conformational isomers (the same molecule)
        e. structural isomers

4.27

        a. Cis–trans isomers not possible
        b. Cis–trans isomers not possible
        c.

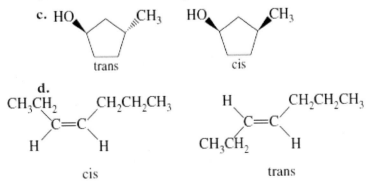

        trans   cis

        d.

        cis   trans

Copyright © 2014 Pearson Education, Inc.
54

4.29

a. 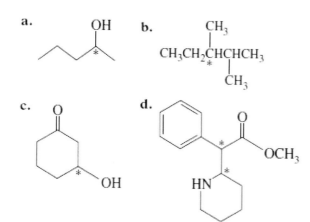 OH

b.

CH₃
|
CH₃CH₂CHCHCH₃
      *
      |
      CH₃

c.

O

OH

d.

O

OCH₃

HN

Additional Problems

4.31   a. inorganic   b. organic   c. organic   d. inorganic
        e. inorganic   f. organic

4.33   Hydrocarbons are organic compounds composed only of carbon and hydrogen. Saturated refers
        to the fact that each carbon is bonded to the maximum number of hydrogens.

4.35

C₅H₁₀

H H
 \|
  C
 / \
H—C   C—H
H| |H
  C—C
 /| |\
H H H H

4.37   Decane

4.39

 a. CH₃CH₂CH₂CH₃   b.   CH₃CH₂CH₂OH
        c. CH₃CH₂CH₂CH₂CH₂CH₂CH₃

4.41

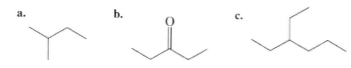

a.   b.   O   c.

4.43

        a. CH₃CH(OH)CH₂CH₃
        b. CH₃CH₂CH₂Br
        c. CH₃CH₂COOH

Copyright © 2014 Pearson Education, Inc.

4.45

a.

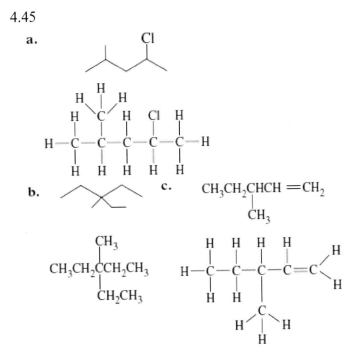

b. 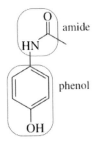   c.   CH₃CH₂CHCH ═CH₂
                              |
                              CH₃

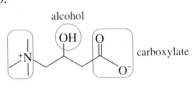

4.47   An unsaturated fatty acid contains one or more carbon–carbon double bonds, but a saturated fatty acid contains no double bonds.

4.49

a.

amide

HN

phenol

OH

Acetaminophen

b.

alcohol

OH

carboxylate

amine   Carnitine

4.51   A = quaternary ammonium   B = sulfide (thioether), C = ether,   D = amine

Copyright © 2014 Pearson Education, Inc.

4.53

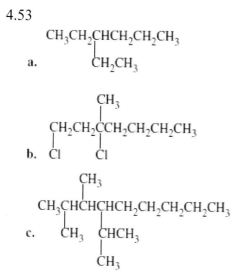

a. CH₃CH₂CHCH₂CH₂CH₃
           |
           CH₂CH₃

b.
                CH₃
                |
   CH₂CH₂CCH₂CH₂CH₂CH₃
   | |
   Cl   Cl

c.
              CH₃
              |
   CH₃CHCHCHCH₂CH₂CH₂CH₃
      | |
      CH₃ CHCH₃
              |
              CH₃

4.55

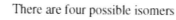

```
         :F: :Cl:
          |   |
    :F—C—C—Br:
          |   |
         :F:  H
```

4.57

CH₃CH₂CH₂CH₂CH₂CH₃   CH₃CH₂   CH₃CH₂
                         | |
                         CH₂CH₂CH₂CH₃   CH₂CH₂CH₂
                                                    |
                                                    CH₃

4.59

There are four possible isomers

Cyclopentane   Methylcyclobutane   1,2-Dimethylcylopropane

1,1-Dimethylcylopropane

Copyright © 2014 Pearson Education, Inc.

4.61

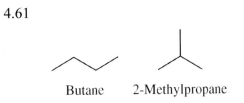

Butane   2-Methylpropane

4.63

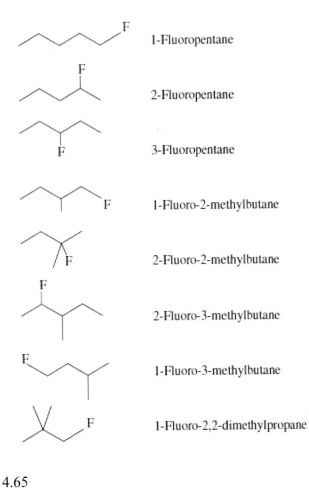

1-Fluoropentane

2-Fluoropentane

3-Fluoropentane

1-Fluoro-2-methylbutane

2-Fluoro-2-methylbutane

2-Fluoro-3-methylbutane

1-Fluoro-3-methylbutane

1-Fluoro-2,2-dimethylpropane

4.65

a.

cis   trans

b.

cis   trans

Copyright © 2014 Pearson Education, Inc.
58

4.67

a. no cis–trans isomer

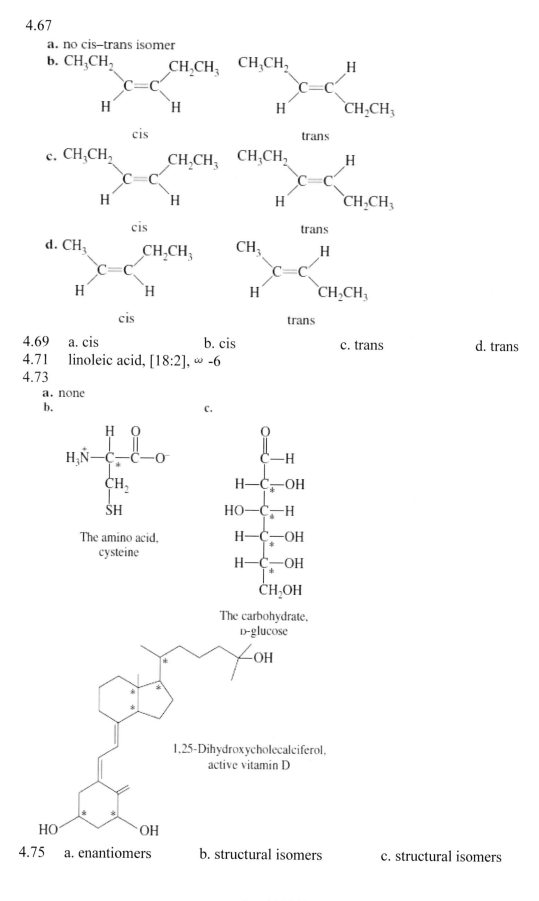

b.

CH₃CH₂ ... CH₂CH₃ / C=C / H ... H — cis

CH₃CH₂ ... H / C=C / H ... CH₂CH₃ — trans

c.

CH₃CH₂ ... CH₂CH₃ / C=C / H ... H — cis

CH₃CH₂ ... H / C=C / H ... CH₂CH₃ — trans

d.

CH₃ ... CH₂CH₃ / C=C / H ... H — cis

CH₃ ... H / C=C / H ... CH₂CH₃ — trans

4.69   a. cis   b. cis   c. trans   d. trans

4.71   linoleic acid, [18:2], ω -6

4.73

a. none

b.

The amino acid, cysteine

c.

The carbohydrate, D-glucose

1,25-Dihydroxycholecalciferol, active vitamin D

4.75   a. enantiomers   b. structural isomers   c. structural isomers

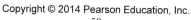
Copyright © 2014 Pearson Education, Inc.

Challenge Problems

4.77

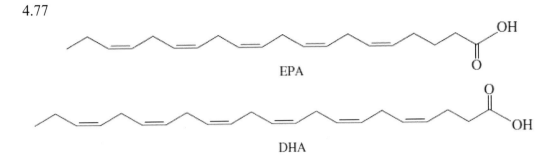

EPA

DHA

Copyright © 2014 Pearson Education, Inc.

Chemical Reactions

5.1 Thermodynamics

Learning Objectives

Upon completion of this material, a student should be able to do the following:
   A. Define the following key terms:

thermodynamics	**free energy change (ΔG)**
reaction kinetics	**spontaneous processes**
exothermic reactions	**nonspontaneous processes**
endothermic reactions	**calorimeter**
free energy (G)	**combustion**

   B. Draw reaction energy diagrams for exergonic and endergonic reactions.
   C. Predict spontaneity of a reaction based the ΔG value.
   D. Describe how a calorimeter works.
   E. Calculate the energy content in foods from its nutrient molecules.

5.2   Chemical Reactions: Kinetics

Learning Objectives

Upon completion of this material, a student should be able to do the following:
   A. Define the following key terms:

rate of reaction	**biochemical reactions**
catalyst	**active site**
enzyme	

   B. Predict relative activation energies and speed of reactions using a reaction energy diagram.
   C. Determine the effect that temperature, amount of reactants, and a catalyst have of the rate of a reaction.
   D. Describe how an enzyme catalyses a biochemical reaction.

5.3   Overview of Chemical Reactions

Learning Objectives

Upon completion of this material, a student should be able to do the following:
   A. Define the following key terms:

synthesis reactions	**chemical equilibrium**
decomposition reactions	**irreversible reactions**
exchange reactions	**mechanism**
reversible reactions	**glycolysis**

   B. Classify reactions as synthesis, decomposition, or exchange reactions.
   C. Distinguish reversible and irreversible reactions.

Copyright © 2014 Pearson Education, Inc.

D. Predict the products and balance the chemical equation for a hydrocarbon undergoing combustion.

E. Contrast a general chemical equation and an organic chemical equation.

5.4 Oxidation and Reduction

Learning Objectives

Upon completion of this material, a student should be able to do the following:

A. Define the following key terms:

oxidation   oxidizing agent

reduction   cellular respiration

reducing agent

B. Identify the substance oxidized and the substance reduced in an inorganic oxidation–reduction reaction.

C. Identify the substance oxidized and the substance reduced in an organic oxidation–reduction reaction.

5.5 Organic Reactions: Condensation and Hydrolysis

Learning Objectives

Upon completion of this material, a student should be able to do the following:

A. Define the following key terms:

condensation   carboxylation

dehydration   phosphorylation

hydrolysis   dephosphorylation

B. Predict the products of an organic condensation reaction.

C. Predict the products of an organic hydrolysis reaction.

5.6 Organic Addition Reactions to Alkenes

Learning Objectives

Upon completion of this material, a student should be able to do the following:

A. Define the following key terms:

addition to alkenes   Markovnikov's rule

B. Predict the products of a hydrogenation reaction, an addition reaction of an alkene.

C. Predict the products of a hydration reaction, an addition reaction of an alkene.

Copyright © 2014 Pearson Education, Inc.

Practice Test for Chapter 5

1. Consider the following reaction energy diagram.

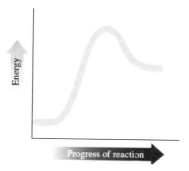

Which of the following correlates with the reaction represented in the diagram?

A. positive ΔG
B. nonspontaneous
C. endergonic
D. all of the above

2. Consider the following reaction energy diagram.

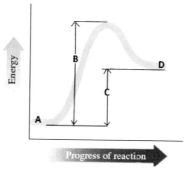

Which letter represents the energy of reaction?
A. A
B. B
C. C
D. D

Copyright © 2014 Pearson Education, Inc.

3.   Consider the following reaction energy diagram.

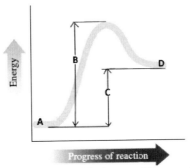

Which quantity shown in the diagram would change if a catalyst were added to this reaction?

A. A
B. B
C. C
D. D
E. Both B and C would change.

4. Consider the following reaction energy diagram.

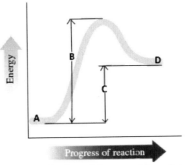

For this reaction, what approximate relationship would exist between the energy of the reaction and the activation energy?

A. The energy of reaction and the activation energy are about the same.
B. The energy of reaction is about double the activation energy.
C. The energy of reaction is about half the activation energy.
D. The energy of reaction is about one quarter the activation energy.
E. The energy of reaction is about four times the activation energy.

5. Butter turning rancid is an example of an oxidation reaction with atmospheric oxygen. Which of the following would not decrease the rate of this reaction?
A. placing the butter in a closed container
B. storing the butter in an atmosphere of nitrogen
C. putting the butter in a refrigerator at 40 °F
D. All of these will decrease the rate at which rancidity occurs.

Copyright © 2014 Pearson Education, Inc.

6. Which of the following is the most likely to represent an equilibrium reaction?
   A. Snow melting when the air temperature is 32 °F.
   B. Butane from a disposable lighter burning.
   C. Nitroglycerin causing an explosion.
   D. Digestion of protein found in food.
   E. None of these is an equilibrium reaction.

7. The following symbols appear in chemical equations. Which indicates that the reaction represents an equilibrium?
   A.

   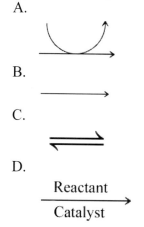

   B.

   C.

   D.
      Reactant
      ⟶
      Catalyst

8. Consider the following representation of reaction conditions.

   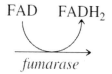

   In what class of reaction would this type of symbolism be found?
   A. inorganic reaction
   B. organic reaction
   C. biochemical reaction
   D. exchange reaction
   E. decomposition reaction

9. Consider the following set of reaction conditions.

   Which of the following is correct about the accompanying reaction?
   A. The reaction is endergonic.
   B. The reaction would be enzyme catalyzed.
   C. The reaction would occur in a biochemical pathway.
   D. All of the above are correct.

Copyright © 2014 Pearson Education, Inc.

10. Consider the following reaction energy diagram.

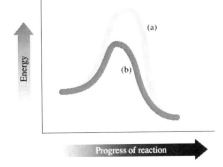

Based on this diagram, which of the following statements is (are) correct?
A. Curve b represents the uncatalyzed pathway.
B. Curve a represents the catalyzed pathway.
C. An exothermic reaction is represented.
D. Curve b has the higher activation energy.
E. None of the statements are correct.

11. Consider the following two reaction energy diagrams.

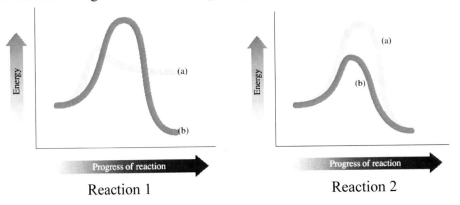

Reaction 1   Reaction 2

If in both diagrams the energy scale is the **same**, which is the fastest exothermic reaction?
A. Reaction 1, pathway a
B. Reaction 1, pathway b
C. Reaction 2, pathway a
D. Reaction 2, pathway b

12. A can of microwaveable soup has the following nutritional facts on the label.
22 g fat
30 g carbohydrate
6 g protein
If there are two servings per container, to two significant figures, approximately how many Calories are there in one serving?
A. 170 Cal
B. 340 Cal
C. 110 Cal

Copyright © 2014 Pearson Education, Inc.

D. 260 Cal

13. Which of the following correctly ranks the type of nutrient molecules in order of increasing energy content?
A. protein < fat < carbohydrate
B. protein ≈ carbohydrate < fat
C. fat < protein ≈ carbohydrate
D. carbohydrate < protein < fat
E. carbohydrate ≈ fat < protein

14. What type of reaction is represented by the following chemical equation?

$$2KClO_3(s) \rightarrow 2KCl(s) + 3O_2(g)$$

A. exchange reaction
B. equilibrium reaction
C. decomposition reaction
D. synthesis reaction

15. The following is graphic representation for the reaction of a polymer and its monomers.

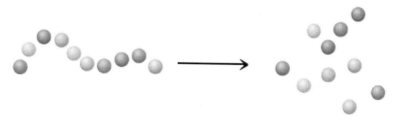

How would this reaction be classified?
A. exchange reaction
B. equilibrium reaction
C. decomposition reaction
D. synthesis reaction

16. What is (are) the most likely product(s) of the reaction shown below?
$$Ca(s) + Cl_2(g) \rightarrow$$
A. $CaCl_2$
B. $CaCl$
C. $CaI + C$
D. $Ca + I_2 + C$

17. What is (are) the most likely product(s) of the reaction shown below?
$$BaF_2(aq) + K_2S(aq) \rightarrow$$
A. $BaS + 2 KF$
B. $F_2S + K_2Ba$
C. $BaKS + F_2$
D. $K_2F_2S + Ba$

Copyright © 2014 Pearson Education, Inc.

18. In the complete combustion of pentane (C₅H₁₂), how many molecules of oxygen are used?

    A. 6
    B. 5
    C. 8
    D. 3
    E. 2

19. An oxidation reaction may involve:
    A. loss of electrons
    B. loss of hydrogen
    C. gain of oxygen
    D. A, B, and C
    E. neither A, nor B, nor C.

20. Consider the following reaction.

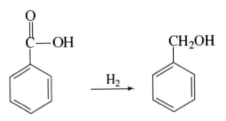

    Which of the following correctly describes this reaction?
    A. The organic reactant is oxidized.
    B. The organic reactant is the reducing agent.
    C. The product of the reaction is an alcohol
    D. All of the above characterize this reaction.

21. Which of the following changes would represent a reduction?
    A. Mg forms MgO in air.
    B. Cu^{2+} ions form Cu_2O during a laboratory sugar test.
    C. The biomolecule NADH forms NAD^+.
    D. CH_3OH forms HCOOH.
    E. All represent reductions.

Copyright © 2014 Pearson Education, Inc.

22. What type of reaction is shown in the following equation?

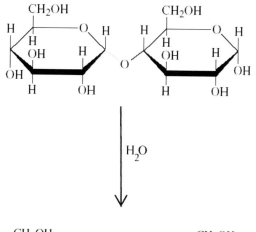

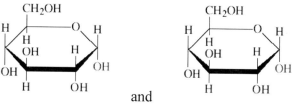

and

A. condensation
B. hydrolysis
C. oxidation
D. reduction

23. What type of reaction is shown in the following equation?

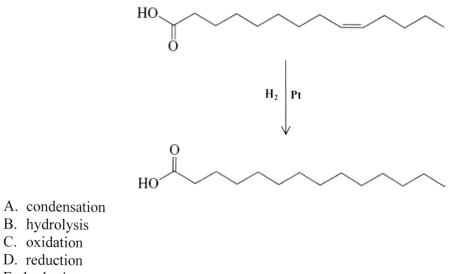

A. condensation
B. hydrolysis
C. oxidation
D. reduction
E. hydration

Copyright © 2014 Pearson Education, Inc.

24. What type of reaction is shown in the following equation?

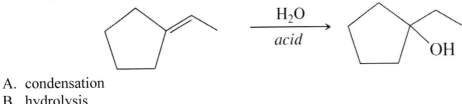

A. condensation
B. hydrolysis
C. oxidation
D. reduction
E. hydration

25. What type of reaction is shown in the following equation?

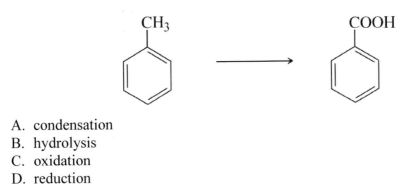

A. condensation
B. hydrolysis
C. oxidation
D. reduction
E. hydration

26. Which of the following represents the appropriate reaction conditions to carry out the following conversion?

$$CH_3CH_2CH=CHCH_3 \longrightarrow CH_3CH_2CH_2CH_2CH_3$$

A. $\xrightarrow{H_2}$

B. $\xrightarrow{H_2O}$

C. $\xrightarrow{\text{oxidation}}$

D. $\xrightarrow{\text{NADH}}$

27. Which of the following statements characterizes the reaction shown below?

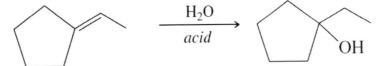

A. The product shown is not that predicted by Markovnikov's rule.

Copyright © 2014 Pearson Education, Inc.

B. The reaction is an example of an addition reaction.
C. The reverse reaction is a hydrolysis reaction.
D. The water would serve as a catalyst.
E. All of the above characterize this reaction.

28. Which of the following is the major organic product of the following reaction?

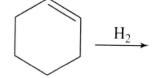

A.

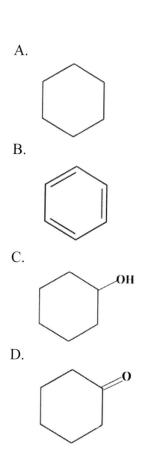

B.

C.

D.

29. The following structure represents the product of a reaction with hydrogen in the presences of a Pt catalyst.

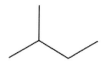

Copyright © 2014 Pearson Education, Inc.

Which of the following could be used as a reactant in this reaction?

A.

B.

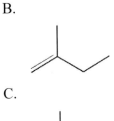

C.

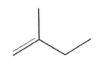

D. either A or C
E. A, B, or C

30. The following structure represents the product of a reaction with water in the presence of an acid catalyst.

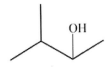

Which of the following could be used as a reactant in this reaction?

A.

B.

C.

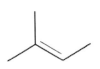

D. either A or C
E. A, B, or C

Copyright © 2014 Pearson Education, Inc.

Answers

1. D 2. C 3. B 4. C 5. D 6. A 7. C 8. C 9. D 10. C
11. D 12. A 13. B 14. C 15. C 16. A 17. A 18. C 19. D 20. C
21. B 22. B 23. D 24. E 25. C 26. A 27. B 28. A 29. E 30. A

Copyright © 2014 Pearson Education, Inc.

Practice Problems

5.1   Cold. If the reaction is endothermic, it absorbs heat from its surroundings, which cools the reaction. This reaction would have a $-\Delta G$, because it is spontaneous.

5.3   a. exothermic   b. endothermic

5.5   a. spontaneous   b. spontaneous

5.7   135 calories (rounds to 100 calories -1 significant figure)

5.9   a. Increasing the temperature increases the rate by increasing the velocity of the reactant molecules. The faster they move, the more likely they are to collide and react.
        b. Increasing the concentration of the reactant increases the likelihood of collision and increases the rate of the reaction.

5.11   The concentration of reactants decreases as the reaction progresses, so the rate slows.

5.13   a. decrease   b. increase   c. increase

5.15   a. exchange   b. decomposition   c. synthesis

5.17   a. irreversible   b. irreversible   c. reversible

5.19 $2C_2H6(g) + 7O_2(g) \rightarrow 4CO_2(g) + 6H_2O(g)$

5.21   In both organic and biochemical reactions, the structure of the organic molecule is shown. The yields arrow shows reversibility for both. Also, changes in the structure of functional groups are shown.

5.23   a. reduction   b. oxidation   c. oxidation

5.25   Hydrogen is oxidized, and oxygen is reduced.

5.27

$$H_3C-\overset{\overset{\displaystyle O}{\|}}{C}-OCH_2CH_3$$

5.29

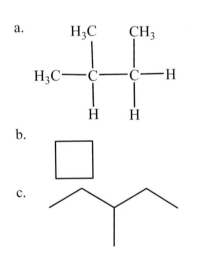

a.

b.

c.

5.31

a.

b.

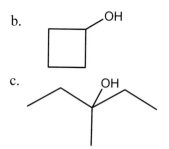

c.

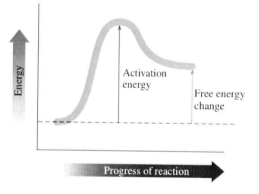

Additional Problems

5.33   a. exothermic   b. reactants   c. Yes, it is spontaneous.

d.

5.35   An exothermic reaction with a low activation energy occurs faster. The low activation energy allows the reactants to react more quickly.

5.37   a. Curve (a) is faster, curve (b) is slower.
       b. Curve (a) is endothermic, curve (b) is exothermic.
       c. (b) is a spontaneous reaction.
       d. (a) has a positive ΔG.

5.39

5.41 780 calories (rounds to 800 calories)
5.42 106 calories (rounds to 100 calories)

5.43   a. $Mg(s) + Cl_2(g) \longrightarrow MgCl_2(s)$

       b. $2HI(g) \longrightarrow H_2(g) + I_2(g)$

       c. $Ca(s) + Zn(NO_3)_2(aq) \longrightarrow Zn(s) + Ca(NO_3)_2(aq)$

       d. $K_2S(aq) + Pb(NO_3)_2(aq) \longrightarrow PbS(s) + 2KNO_3(aq)$

Copyright © 2014 Pearson Education, Inc.

5.45

a.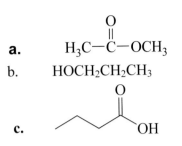

b. $HOCH_2CH_2CH_3$

c.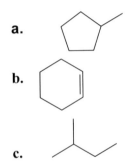

5.47   a. condensation   b. oxidation   c. oxidation
5.49 $C_5H_{12}(g) + 8 O_2(g) \rightarrow 5 CO_2(g) + 6 H_2O(g)$
5.51   a. Magnesium is oxidized, and iron is reduced.
        b. Aluminum is oxidized, oxygen is reduced.
        c. Bromide is oxidized, silver is reduced.

5.53

a.

b.

c.

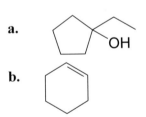

5.55

a.

b.

c. $CH_2 = CHCH_3$

5.57   a. The forward reaction (glucose to glycogen) is condensation; the reverse reaction (glycogen to glucose) is hydrolysis.
        b. On a low carbohydrate diet, glycogen stores are depleted and a significant amount of water is also lost. So the weight lost is not fat but rather glycogen and water.

Copyright © 2014 Pearson Education, Inc.

Challenge Problems

5.59   60 days

5.61

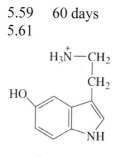

Serotonin

Copyright © 2014 Pearson Education, Inc.

Carbohydrates: Life's Sweet Molecules

6.1 Classes of Carbohydrates

Learning Objectives

Upon completion of this material, a student should be able to do the following:

    A. Define the following key terms:

        carbohydrates   **soluble fiber**

        monosaccharides   **insoluble fiber**

        disaccharides   **polysaccharides**

        oligosaccharides

    B. Classify carbohydrates as mono-, di-, oligo-, or polysaccharides.

    C. Distinguish soluble and insoluble fibers.

6.2   Functional Groups in Monosaccharides

Learning Objectives

Upon completion of this material, a student should be able to do the following:

    A. Define the following key terms:

        alcohol   **ketone**

        primary (1°) alcohol   **aldose**

        secondary (2°) alcohol   **ketose**

        tertiary (3°) alcohol   **glucose**

        carbonyl   **fructose**

        aldehyde

    B. Distinguish primary, secondary, and tertiary alcohols.

    C. Recognize and draw the functional groups alcohol, aldehyde, and ketone.

6.3   Stereochemistry in Monosaccharides

Learning Objectives

Upon completion of this material, a student should be able to do the following:

    A. Define the following key terms:

        Fischer projection   **glycolysis**

        D-sugar   **galactose**

        L-sugar   **epimers**

        diastereomer   **mannose**

Copyright © 2014 Pearson Education, Inc.

B. Distinguish D- and L- stereoisomers of monosaccharides.
C. Draw Fischer projections of linear monosaccharides.
D. Define enantiomer, epimer, and diastereomer.
E. Draw enantiomers and diastereomers of linear monosaccharides.
F. Characterize common monosaccharides.

6.4   Reactions of Monosaccharides

Learning Objectives

Upon completion of this material, a student should be able to do the following:
A. Define the following key terms:

hemiacetal	**furanose**
anomers	**Benedict's test**
anomeric carbon	**reducing sugar**
pyranose	

B. Draw cyclic α and β anomers from linear monosaccharide structures.
C. Draw oxidation and reduction products of aldoses.

6.5   Disaccharides

Learning Objectives

Upon completion of this material, a student should be able to do the following:
A. Define the following key terms:

glycosidic bond	**lactose**
glycoside	**sucrose**
maltose	

B. Locate and name glycosidic bonds in disaccharides.
C. Distinguish condensation and hydrolysis reactions of simple sugars.
D. Characterize common dissacharides.

6.6   Polysaccharides

Learning Objectives

Upon completion of this material, a student should be able to do the following:
A. Define the following key terms:

storage polysaccharides	**glycogen**
structural polysaccharides	**cellulose**
amylose	**chitin**
amylopectin	

B. Identify polysaccharides by glycosidic bond and sugar subunit.

Copyright © 2014 Pearson Education, Inc.

6.7 Carbohydrates and Blood

Learning Objectives

Upon completion of this material, a student should be able to do the following:

A. Define the following key terms:

fucose   **glycoaminoglycans**

heparin

B. Predict ABO compatibility.

C. Describe the structure and role of heparin.

Copyright © 2014 Pearson Education, Inc.

Practice Test for Chapter 5

1. The following structure could be the representation of a(n):

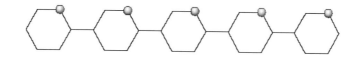

    A. monosaccharide.
    B. disaccharide.
    C. oligosaccharide.
    D. polysaccharide.

2. The molecule represents epinephrine (adrenaline). Epinephrine contains:

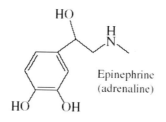

Epinephrine
(adrenaline)

    A. one primary alcohol substituent.
    B. one secondary alcohol substituent.
    C. three secondary alcohol substituents.
    D. one tertiary alcohol substituent.

Copyright © 2014 Pearson Education, Inc.

3. Which of the following contains a ketone functional group?

A.

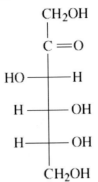

CH$_2$OH
|
C$=$O

HO —|— H

H —|— OH

H —|— OH

CH$_2$OH

B.

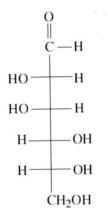

O
‖
C — H

HO —|— H

HO —|— H

H —|— OH

H —|— OH

CH$_2$OH

C.

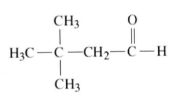

$$CH_3$$
$$|$$
$$H_3C - C - CH_2 - \overset{\displaystyle O}{\overset{\|}{C}} - H$$
$$|$$
$$CH_3$$

D.

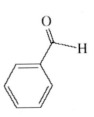

Copyright © 2014 Pearson Education, Inc.

4. How many chiral centers are in the following monosaccharide?

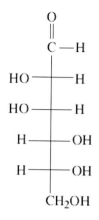

A. 2
B. 3
C. 4
D. 5
E. 6

5. In a Fischer projection, the difference between the D-sugar and the L-sugar is:
   A. the presence of an aldehyde versus a ketone functional group.
   B. the number of chiral centers.
   C. the position of the —OH group closest to the carbonyl atom.
   D. the position of the —OH group furthest from the carbonyl carbon atom.
   E. the position of the —OH group on the chiral center furthest from the carbonyl carbon atom.

Copyright © 2014 Pearson Education, Inc.

6. Which of the following would be classified as an aldohexose?

A.

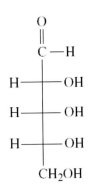

B.

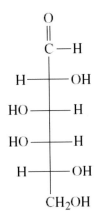

C.

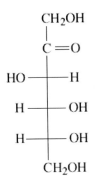

D. None of these is an aldohexose.

Copyright © 2014 Pearson Education, Inc.

7. Which of the following represents the L-sugar of an aldose?

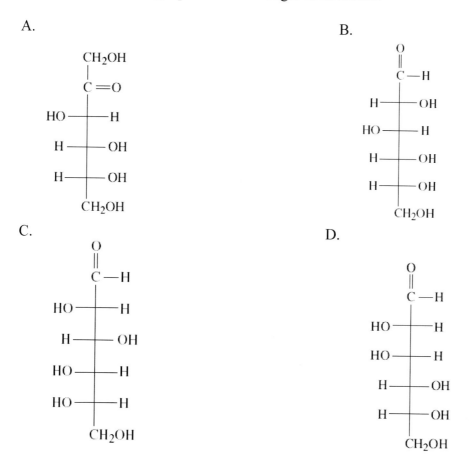

A.

CH₂OH
|
C=O
HO——H
H——OH
H——OH
CH₂OH

B.

O
||
C—H
H——OH
HO——H
H——OH
H——OH
CH₂OH

C.

O
||
C—H
HO——H
H——OH
HO——H
HO——H
CH₂OH

D.

O
||
C—H
HO——H
HO——H
H——OH
H——OH
CH₂OH

E. None of these is the L-sugar of an aldose.

Copyright © 2014 Pearson Education, Inc.

8. Which of the following represents an enantiomer of:

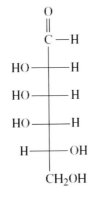

A.

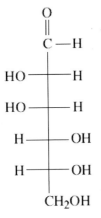

B.

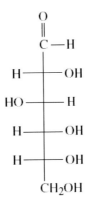

C.

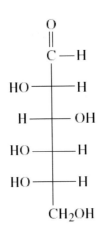

D.

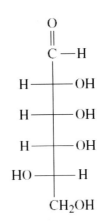

E. None of these is an enantiomer.

Copyright © 2014 Pearson Education, Inc.

9. Which of the following represents an epimer of:

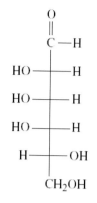

A.

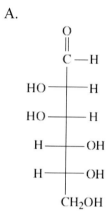

B.

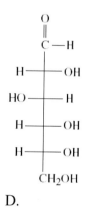

C.

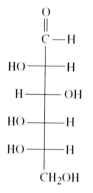

D.

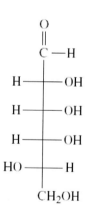

E. None of these is an epimer.

Copyright © 2014 Pearson Education, Inc.

10. Which of the following represents a diastereomer of:

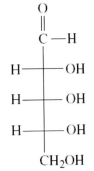

A.

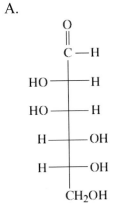

B.

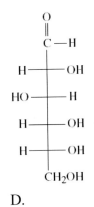

C.

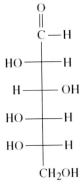

D.

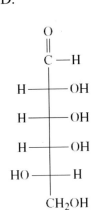

E. None of these is a diastereomer.

Copyright © 2014 Pearson Education, Inc.

11. What is the product formed when the following monosaccharide is treated with Benedict's reagent?

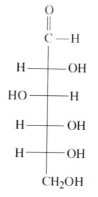

A.

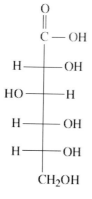

B.

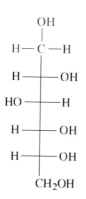

C.

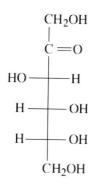

D. The given monosaccharide is a nonreducing sugar.

Copyright © 2014 Pearson Education, Inc.

12. The following compound would be designated as:

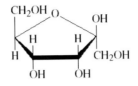

   A. pyranose.
   B. furanose.
   C. nonreducing sugar.
   D. both A and C
   E. both B and C

13. Which of the following would be designated as an α anomer?
   A.

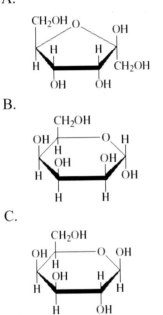

   D. All are β anomers.

14. When an open chain monosaccharide forms a ring, a(n) _____ functional group is formed. Which of the following should be used in the blank?
   A. aldehyde
   B. ketone
   C. tertiary alcohol
   D. hemiacetal

Copyright © 2014 Pearson Education, Inc.

15. What is the correct ring structure for the α anomer in pyranose ring form for the following monosaccharide?

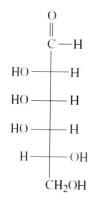

A.

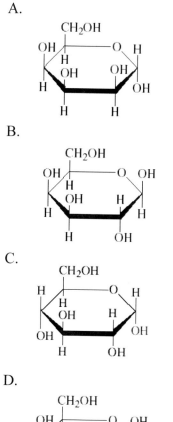

B.

CH₂OH

C.

CH₂OH

D.

CH₂OH

Copyright © 2014 Pearson Education, Inc.

16. What will be the difference in the ring forms of the following two monosaccharides?

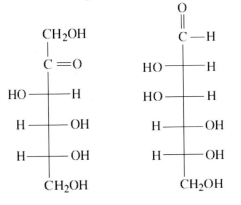

A. the number of atoms in the ring
B. the number of anomers possible
C. the placement of the —OH on C1 relative to the ring (above versus below)
D. the product being a reducing or nonreducing sugar

17. Which of the following in **not** a characteristic of fructose?
   A. also called fruit sugar
   B. a component of sucrose
   C. epimer of glucose
   D. sweetest monosaccharide

18. A monosaccharide is a component of both lactose and sucrose and is commonly known as blood sugar. This monosaccharide is:
   A. mannose.
   B. galactose.
   C. glucose.
   D. ribose.

19. Consider the following structure.

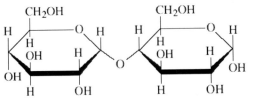

   This structure represents:
   A. a nonreducing oligosaccharide.
   B. a reducing disaccharide.
   C. the product of a hydrolysis reaction.
   D. both B and C

Copyright © 2014 Pearson Education, Inc.

20. The following reaction symbolically represents a:

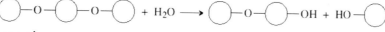

A. hemiacetal reaction.
B. dehydration reaction.
C. condensation reaction.
D. hydrolysis reaction.

21. Which of the following does **not** contain a glycosidic bond?
   A.

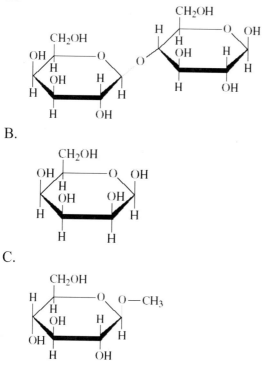

   B.

   C.

   D. B and C
   E. A, B, and C contain glycosidic bonds.

22. Determine the type of glycosidic bond in the following structure.

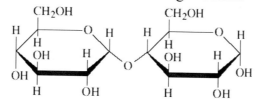

   A. $\alpha(1\rightarrow4)$
   B. $\beta(1\rightarrow4)$
   C. $\alpha(1\rightarrow6)$
   D. $\beta(1\rightarrow6)$

Copyright © 2014 Pearson Education, Inc.

23. The product of the following reaction would be:

glucose + galactose $\longrightarrow$

A. maltose.
B. lactose.
C. sucrose.
D. amylose.

24. Which of the following describes sucrose?
A. a reducing sugar
B. undergoes hydrolysis to produce one type of monosaccharide
C. most abundant disaccharide in nature
D. found in mammalian milk
E. All of the above describe sucrose.

25. The product(s) of the hydrolysis reaction of the following sugar would be:

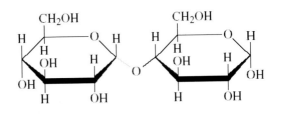

A. 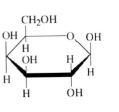 and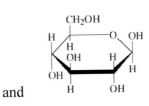

B. 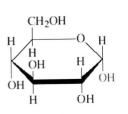 and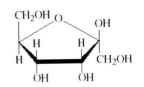

C.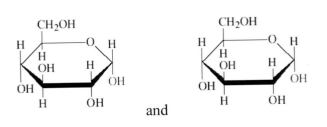

D. Anomers cannot be determined from given information.

Copyright © 2014 Pearson Education, Inc.

26. Consider the segment of the following polysaccharide:

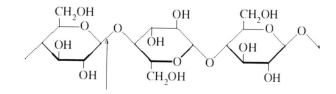

The arrow in the figure represents a(n):
A. $\alpha(1\rightarrow4)$ bond.
B. $\beta(1\rightarrow4)$ bond.
C. $\alpha(1\rightarrow6)$ bond.
D. $\beta(1\rightarrow6)$ bond.

27. Which of the following is an example of a structural polysaccharide?
A. amylose.
B. amylopectin.
C. glycogen.
D. chitin.

28. Consider a storage polysaccharide that contains $\alpha(1\rightarrow4)$ bonds, is found in starch, and has branches about every 25 glucose units. This polysaccharide is:
A. amylose.
B. amylopectin.
C. glycogen.
D. cellulose.

29. Type O blood is available for transfusion. Recipients with the following blood types need a transfusion. Which can receive type O blood?
A. A
B. B
C. AB
D. A, B, and AB
E. Only recipients with type O blood would be compatible.

30. Consider the following, which represents the "backbone" of the blood type groups.

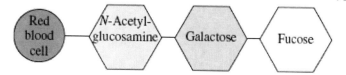

Groups bonded to which unit of the backbone distinguish the various blood types?
A. red blood cell
B. N-acetylglucosamine
C. galactose
D. fucose

Copyright © 2014 Pearson Education, Inc.
95

Answers

1. C 2. B 3. A 4. C 5. E 6. B 7. C 8. D 9. A 10. E
11. A 12. B 13. B 14. D 15. A 16. A 17. D 18. C 19. B 20. D
21. B 22. A 23. B 24. C 25. C 26. B 27. D 28. B 29. D 30. C

Copyright © 2014 Pearson Education, Inc.

Chapter 6 – Solutions to Odd-Numbered Problems

Practice Problems

6.1   a. polysaccharide   b. oligosaccharide   c. disaccharide

6.3   a. secondary   b. primary   c. tertiary   d. secondary

6.5   a. ketone   b. ketone   c. aldehyde

6.7   a. D-isomer   b. L-isomer   c. D-isomer

6.9

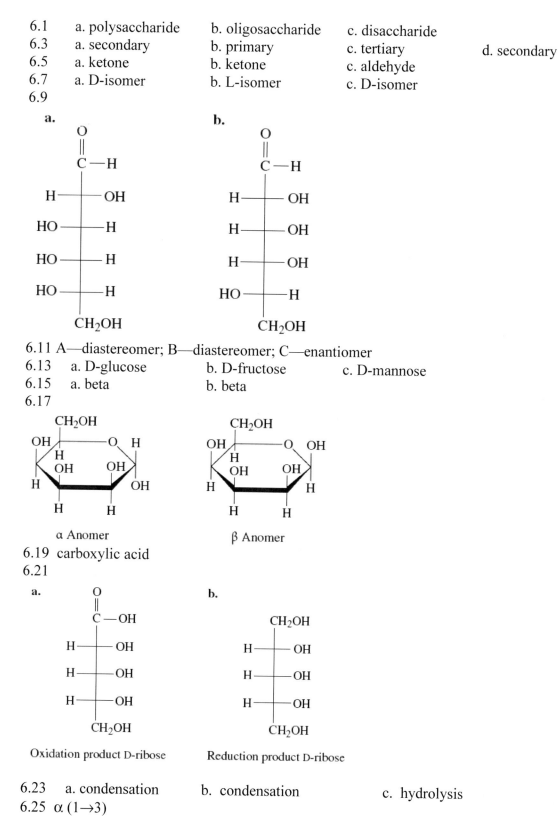

6.11 A—diastereomer; B—diastereomer; C—enantiomer

6.13   a. D-glucose   b. D-fructose   c. D-mannose

6.15   a. beta   b. beta

6.17

α Anomer   β Anomer

6.19 carboxylic acid

6.21

Oxidation product D-ribose   Reduction product D-ribose

6.23   a. condensation   b. condensation   c. hydrolysis

6.25 $\alpha\,(1{\rightarrow}3)$

Copyright © 2014 Pearson Education, Inc.

6.27

    a. bond is $\beta(1 \rightarrow 4)$

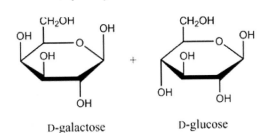

           D-galactose   D-glucose

  b. bond is $\alpha(1 \rightarrow 4)$

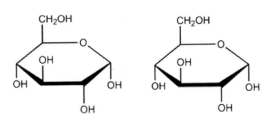

       Both monosaccharides are D-glucose

6.29   a. sucrose   b. lactose   c. maltose   d. lactose

6.31   a. Both contain $\alpha(1 \rightarrow 4)$ glycosidic bonds and only D-glucose. Amylopectin also contains branching $\alpha(1 \rightarrow 6)$.

        b. Both contain $\alpha(1 \rightarrow 4)$ glycosidic bonds and branching $\alpha(1 \rightarrow 6)$ and only D-glucose. Branching occurs more often in glycogen than amylopectin.

6.33   a. cellulose, chitin

        b. amylose, amylopectin

        c. amylose

        d. glycogen

6.35   a. No. A person with type B blood can only receive type B or O blood.

        b. No. A person with type B blood can only receive type B or O blood.

Additional Problems

6.37 $C_4H_8O_4$

6.39   An oligosaccharide is smaller, containing between three and nine monosaccharide units, while a polysaccharide contains 10 or more monosaccharide units.

6.41   Soluble fibers mix with water and form a gel-like substance that gives a feeling of fullness when eaten.

6.43   aldehyde, hydroxyls (alcohol)

6.45   a. secondary   b. primary   c. secondary

        d. primary   e. secondary

6.47   a. enantiomer   b. epimer

6.49   a. alpha   b. beta

Copyright © 2014 Pearson Education, Inc.

6.51

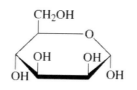
α-D-Mannose

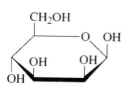
β-D-Mannose

6.53

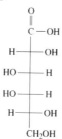

Galactose oxidized at C1

6.55

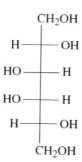

Galactose reduced at C1

6.57   a. yes   b. yes   c. no   d. no

6.59

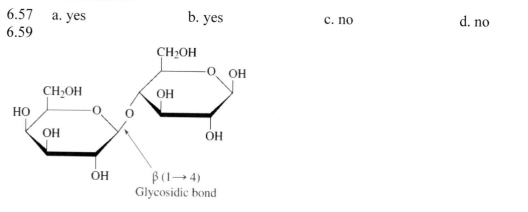

β (1 → 4)
Glycosidic bond

Copyright © 2014 Pearson Education, Inc.

6.61

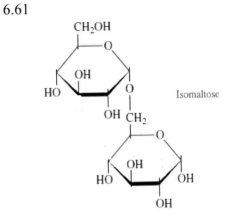

Isomaltose

6.63   a. sucrose   b. glucose   c. L-fucose   d. amylose, amylopectin
        e. chitin
6.65   a. no   b. no

Challenge Problems

6.67
a.

b.

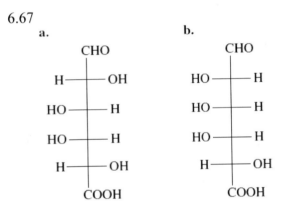

6.69
a.

b.

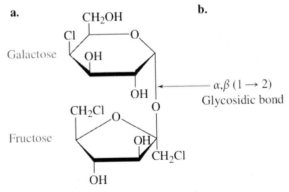

c. Splenda is not a reducing sugar.
6.71   200 Calories; 9%

Copyright © 2014 Pearson Education, Inc.
100

What's the Attraction? State Changes, Solubility, and Cell Membranes

7.1 Types of Attractive Forces

Learning Objectives

Upon completion of this material, a student should be able to do the following:
  A. Define the following key terms:

        attractive force   **hydrogen bonding**
        intermolecular force   **ion-dipole attraction**
        London forces   **ionic attraction**
        induced dipole   **salt**
        permanent dipole   **salt bridge**
        dipole-dipole attraction
  B. Describe five types of attractive forces present in compounds.
  C. Determine the type of intermolecular force based on a chemical formula

7.2 Liquids and Solids: Attractive Forces Are Everywhere

Learning Objectives

Upon completion of this material, a student should be able to do the following:
  A. Define the following key terms:

        changes of state   **sublimation**
        freezing   **deposition**
        melting   **boiling point**
        evaporation   **boiling**
        condensation
  B. Describe the process of boiling.
  C. Predict boiling points for liquids based on the attractive forces present.
  D. Predict melting points for solids based on the attractive forces present.

7.3 Attractive Forces and Solubility

Learning Objectives

Upon completion of this material, a student should be able to do the following:
  A. Define the following key terms:

        solubility   **amphipathic**
        golden rule of solubility   **hydrophobic**
        triglycerides   **hydrophilic**
        esterification   **micelle**
        hydration   **emulsifier**

Copyright © 2014 Pearson Education, Inc.

B. State the golden rule of solubility.
C. Predict the solubility of a molecule in water.
D. Recognize an amphipathic molecule.
E. Define the role of an emulsifier.
F. Draw a fatty acid micelle.

7.4   Gases: Attractive Forces Are Limited

Learning Objectives
Upon completion of this material, a student should be able to do the following:
A. Define the following key terms:

pressure   **pounds per square inch (psi)**
millimeters of mercury (mmHg)

B. Contrast the attractive forces present in a gas with those in a solid or liquid.
C. Define pressure.
D. Apply Boyle's law.
E. Apply Charles's law.

7.5   Dietary Lipids and Trans Fats

Learning Objectives
Upon completion of this material, a student should be able to do the following:
A. Define the following key terms:

fat   **hydrogenation**
oil

B. Distinguish a fat from an oil.
C. Describe the differences in melting points of fats and oils based on their attractive forces.
D. Predict the products of the complete hydrogenation of a triglyceride.

7.6   Attractive Forces and the Cell Membrane

Learning Objectives
Upon completion of this material, a student should be able to do the following:
A. Define the following key terms:

phospholipids   **steroids**
fluid mosaic model

B. Draw a phospholipid bilayer.
C. Describe the structure of a cell membrane.
D. Locate the polar and nonpolar regions of a phospholipid and cholesterol.
E. Classify molecules as steroids based on their structure.

Copyright © 2014 Pearson Education, Inc.

Practice Test for Chapter 7

1. The type of intermolecular force depicted by the arrow in the image is:

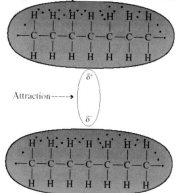

   A. London force.
   B. dipole-dipole attraction.
   C. hydrogen bonding.
   D. ion-dipole attraction.

2. What type(s) of intermolecular attraction would exist between molecules of the following type?

   A. London forces
   B. dipole-dipole attraction
   C. hydrogen bonding
   D. A and B
   E. A, B, and C

3. How many hydrogen bonds could the following molecule form with water?

$$CH_3CHCH_2CH_3$$
$$|$$
$$OH$$

   A. none
   B. 1
   C. 2
   D. 3
   E. 4

Copyright © 2014 Pearson Education, Inc.

4. Identify the strongest attractive force that could exist between the following pair of molecules.

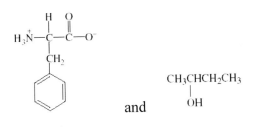

and CH$_3$CHCH$_2$CH$_3$
              |
             OH

A. London forces
B. dipole-dipole attraction
C. hydrogen bonding
D. ion-dipole attraction.

5. Which of the following is a correct description of the manner in which the following molecule would interact with water?

A. This molecule would be a hydrogen bond donor and water a hydrogen bond acceptor.
B. This molecule would be a hydrogen bond acceptor and water a hydrogen bond donor.
C. Both this molecule and water function as hydrogen bond acceptors and donors.
D. This molecule would not hydrogen bond with water.

6. Which of the following pairs of compounds would most likely **not** be soluble in each other?
A. NH$_3$ and CH$_3$OH
B. CH$_3$CH$_2$CH$_2$CH$_3$ and CH$_3$CH$_2$CH$_2$OCH$_2$CH$_3$
C. NH$_4$NO$_3$ and CH$_3$CH$_2$OCH$_2$CH$_3$
D. CH$_3$CH$_2$CH$_2$OH and CH$_3$CH$_2$NH$_2$

7. Compare the solubility of the following substances in water. In which of the pairs of substances would the compound listed on the **left** probably be **more soluble** in water?
A. CH$_3$CH$_2$CH$_2$OH and CH$_3$OH
B. CH$_3$CH$_2$CH$_2$OCH$_2$CH$_3$ and CH$_3$CH$_2$NH$_2$
C. KCl and CO$_2$
D. CH$_3$CH$_2$CH$_3$ and CH$_3$CH$_2$CH$_2$CH$_2$NH$_2$

8. When glycerol and a fatty acid react, the reaction is called a(n):
A. esterification reaction.
B. condensation reaction.
C. hydrogenation reaction.
D. esterification or condensation reaction.
E. triglyceride reaction.

9. The "like" in the expression "like dissolves like" refers to molecular:
A. polarity.
B. size.
C. shape.

Copyright © 2014 Pearson Education, Inc.

D. all of the above

10. Consider the structure of a soap. Which of the following statements is(are) true?
   A. Soaps are amphipathic compounds.
   B. A soap contains a hydrophobic head.
   C. A soap contains a hydrophilic tail.
   D. All of the above statements are true.

11. A soap is able to dissolve nonpolar grease and oil in water by:
   A. forming a micelle around the oil with the soap tails in the interior.
   B. forming a micelle around the oil with the soap heads in the interior.
   C. forming an ion-dipole attraction to the oil.
   D. surrounding the oil with carboxylate group of the soap.

12. In a micelle formed from water and a soap,
   A. hydrophobic heads are oriented toward the surface of the micelle.
   B. hydrophobic tails are oriented toward the surface of the micelle.
   C. hydrophilic heads are oriented toward the surface of the micelle.
   D. hydrophilic tails are oriented toward the surface of the micelle.

13. In "dry" cleaning, an oil stain can be removed by dissolving the oil directly in the liquid being used. Which of the following might make a dry cleaning liquid?
   A. CH_3CH_2OH
   B. $CH_3CH_2CH_2CH_2CH_2CH_3$
   C. $HOCH_2CH_2CH_2CH_2OH$
   D. All would be equally good candidates.

14. Consider the bond formed when a triglyceride is produced. Which of the following general bonding schemes has the same type of bond enclosed in the circle?
   A.

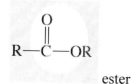

   ester

   B.

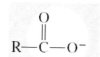

   carboxylate

   C.

   ether

   D.

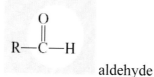

   aldehyde

Copyright © 2014 Pearson Education, Inc.

15. Which of the following substances would have the highest boiling point?
    A. $CH_3CH_2CH_3$
    B. $CH_3CH_2CH_2CH_3$
    C. $CH_3CH_2CH_2CH_2CH_3$
    D. $CH_3CH_2CH_2CH_2CH_2CH_3$

16. Which of the following substances would have the lowest boiling point?
    A.

    B.

    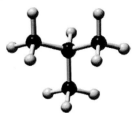

    C.

    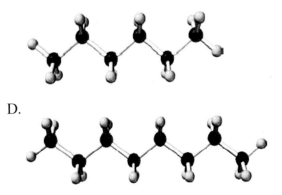

    D.

17. Compare the boiling points of the following substances. In which of the pairs of substances would the compound listed on the **left** have the **higher** boiling point?
    A. $CH_3CH_2CH_2CH_3$ and $CH_3CH_2CH_2OH$
    B. $CH_3CH_2CH_2CH_3$ and $CH_3CH_2NHCH_2CH_3$
    C. $CH_3CH_2CH_2CH_3$ and $CH_3CH_2CH_3$
    D. $CH_3CH_2CH_2CH_3$ and $CH_3CH_2OCH_2CH_3$

18. Which of the following does not characterize both boiling and melting?
    A. heat must be added
    B. phase transition

Copyright © 2014 Pearson Education, Inc.

C. involve the liquid state of matter

D. All characterize melting and boiling.

19. Consider a weather balloon that is filled at sea level where atmospheric pressure is about 760 mmHg and the temperature is 23 °C. This balloon is then transported to the top of a mountain to be released where the pressure is 680 mmHg and the temperature is 15 °C. Which of the following correctly describes how the volume of the balloon will change?

A. The volume will increase due to the pressure change and increase due to the temperature change.

B. The volume will decrease due to the pressure change and decrease due to the temperature change.

C. The volume will increase due to the pressure change and decrease due to the temperature change.

D. The volume will decrease due to the pressure change and increase due to the temperature change.

20. The lung capacity of a adult male diver is 5.40 L at a room pressure of 755 mmHg. If the Diver's lung capacity is reduced to 4.90 L during a dive, what is the pressure in mmHg experienced by the diver?

A. 832 mmHg

B. 685 mmHg

C. 731 mmHg

D. 907 mmHg

21. The 275 mL sample of nitrogen gas is trapped in a container that has a flexible volume at 10.5 °C. What is the temperature of the sample when the volume of the gas is 395 mL? Assume that only the temperature changes.

A. 15.1 K

B. 429 K

C. 208 K

D. 7.31 K

22. Fats:

A. contain mostly unsaturated fatty acids.

B. are carboxylate ions of carboxylic acids.

C. are liquids at room temperature.

D. have melting points near body temperature.

23. Which of the following contains a trans fatty acid?

A.

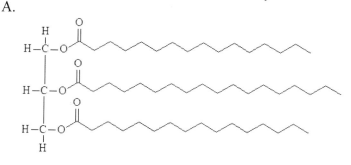

Copyright © 2014 Pearson Education, Inc.

B.

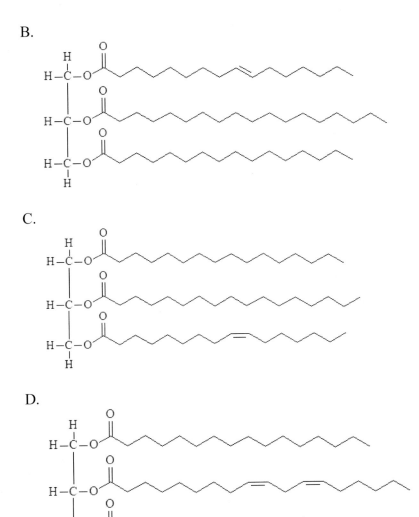

C.

D.

24. How many molecules of hydrogen would be required to completely saturate the following substance?

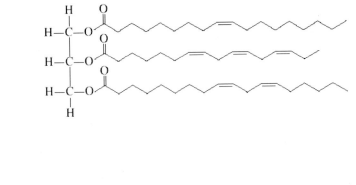

A. 6
B. 9
C. 12
D. 18

Copyright © 2014 Pearson Education, Inc.

25. The following molecule was formed from the reaction of:

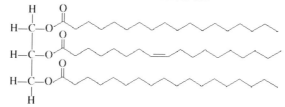

A. glycerol and one saturated fatty acid and one unsaturated fatty acid.
B. glycerol and two different saturated fatty acids and one unsaturated fatty acid.
C. glycerol and one saturated fatty acid and two different unsaturated fatty acids.
D. glycerol and three different saturated fatty acids.

26. An oil was completely hydrogenated with hydrogen gas (H_2) in the presence of a platinum catalyst. The triglycerides comprising the oil contained only one type of fatty acid. If it required 3 molecules of hydrogen for each oil molecule, which of the following fatty acids could be found in the triglycerides?

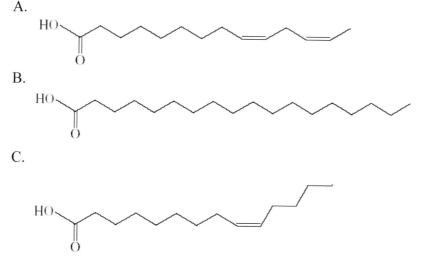

A.

B.

C.

D. The structure of the fatty acid is not represented by any of the above.

27. The following is a cartoon of a soap molecule.

Water would be the most strongly attracted to:
A. the upper charged end.
B. the lower tail.
C. either end.
D. Water is not attracted to the molecule.

Copyright © 2014 Pearson Education, Inc.

28. The primary structural components of cell membranes are:
   A. fatty acids.
   B. triglycerides.
   C. phospholipids.
   D. cholesterol.

29. If the following represents a phospholipid,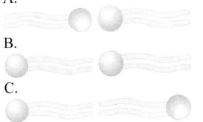

   phospholipids might be arranged **between** the layers of the cell membrane as:
   A.

   B.

   C.

   D. The order of orientation is random.

30. Which of the following modulates the flexibility of a cell membrane?
   A. phospholipids
   B. cholesterol
   C. proteins
   D. triglycerides

Copyright © 2014 Pearson Education, Inc.

Answers

1. A   2. D   3. D   4. D   5. B   6. C   7. C   8. D   9. A   10. A
11. A   12. C   13. B   14. A   15. D   16. B   17. C   18. C   19. C   20. A
21. B   22. D   23. B   24. A   25. A   26. C   27. A   28. C   29. C   30. B

Copyright © 2014 Pearson Education, Inc.

Chapter 7 – Solutions to Odd-Numbered Problems

Practice Problems

7.1   Covalent bonds involve the sharing of electrons between atoms within a molecule, creating a chemical bond. Attractive forces involve attractions between positive and negative areas of different molecules. In attractive forces, electrons are not actually shared between atoms as in a chemical bond.

7.3

Molecule	Hydrogen Bond	Dipole–Dipole	London
CH_3OH	Yes	Yes	Yes
NH_3	Yes	Yes	Yes
CH_3CH_3	No	No	Yes

7.5   The dipole in London forces is temporary; the dipole in dipole–dipole forces is permanent.

7.7   A hydrogen bond acceptor is a pair of electrons on an O, N, or F.

7.9   Polar molecules such as water are not ionic, and pure ionic compounds are attracted more strongly through ionic attractions.

7.11   a. ionic attractions   b. London forces
       c. hydrogen bonding   d. dipole–dipole

7.13   Three hydrogen bonds can be formed.

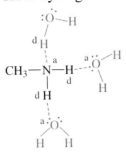

7.15   a. H_2O has the higher boiling point; CH_4, London force; H_2O, H-bonding.
       b. NH_3 has the higher boiling point; NH_3, H-bonding; CO_2, London force.
       c. Acetic acid will have the higher boiling point.

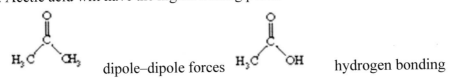

       d. KCl has the higher boiling point; KCl, ionic; CH_3OCH_3, dipole–dipole.

7.17   The highest boiling point is found in the compound with the strongest attractive forces (hydrogen bonding), CH_3OH. The lowest boiling point is found in the compound with the weakest attractive forces (London force), CH_4.

7.19   The alcohol molecule on the right has the highest boiling point because it has the strongest attractive forces present. It is an alcohol and has hydrogen bonding and dipole–dipole attractions. Of the two alkanes, the one with the more branching has the lower boiling point because of its decreased surface area.

7.21   a. insoluble   b. soluble   c. forms a micelle

7.23   a. The fatty acid salt is more soluble in water. The salt can form stronger ion–dipole interactions not available to the fatty acid.

Copyright © 2014 Pearson Education, Inc.

      b. KCl. The ions in KCl readily dissolve in water through strong ion–dipole interactions, while $(CH_3)_3N$ interacts with water via dipole–dipole interactions.

7.25   a. It would look like balloon C.
        b. It would look like balloon B.
        c. It would look like balloon A.

7.27   17 L

7.29   a. It would look most like balloon C.
        b. It would look most like balloon A.
        c. It would look most like balloon B.

7.31   A triglyceride is a dietary lipid containing three fatty acids bonded to a glycerol molecule.

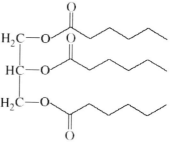

7.33   Even though fats and oils are both triglycerides, the large number of cis double bonds in the hydrocarbon tails of oils prevents them from interacting with each other as much as the saturated tails of the fats. More interactions mean that the tails move more slowly, becoming more like a solid, forming a fat.

7.35   2

7.37   Phospholipids will form two layers called a bilayer with the nonpolar tails of the two layers facing each other and their polar heads facing the water.

Additional Problems

7.39   Both are attractive forces that occur between positive and negative areas. A dipole–dipole attraction occurs between two opposite partial charges. An ion–dipole attraction occurs between opposite ionic charges and partial charges.

7.41

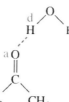

        The molecules can form more than one hydrogen bond. Methylamine can form three hydrogen bonds, acting as an acceptor once (lone pair on nitrogen) and a donor twice.

7.43   a. ionic attraction
        b. London forces
        c. London forces, dipole–dipole attractions, hydrogen bonding
        d. London forces
        e. London forces, dipole–dipole attractions
        f. London forces, dipole–dipole attractions, hydrogen bonding

7.45   An emulsifier can attract both a nonpolar and a polar substance, allowing both substances to be suspended in a mixture.

Copyright © 2014 Pearson Education, Inc.

7.47   a. fatty acid   b. fat   c. CS_2 d. $CH_3CH_2CH_2CH_2CH_2CH_3$

7.49   The stain must be hydrophobic because it is not soluble in water.

7.51   Highest to lowest boiling point: a>c>b>d.
        a. 5 possible H-bonds per molecule (strongest)
        b. polar molecule
        c. 3 possible H-bonds per molecule
        d. nonpolar molecule (weakest)

7.53   Because the hydrocarbon chain is longer in octane (meaning the molecule has a greater surface area), the London forces between molecules are stronger, which makes the boiling point higher for octane.

7.55   A branched alkane has less surface contact with neighboring molecules than does a straight-chain alkane. The attractive forces are therefore stronger between the straight-chain alkane molecules, raising the boiling point.

7.57   Niacin. The niacin overall contains more polar groups versus nonpolar areas than the vitamin A.

7.59   2 cc

7.61   12 L

7.63   10.7 L

7.65   535 °C

7.67

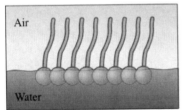

7.69   The soybean oil has more unsaturated fatty acids because it has a lower melting point.

7.71   The phospholipid head group contains many more atoms than the three atoms in the carboxylate head group of a soap.

Challenge Problems

7.73

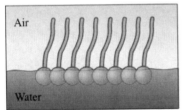

Hydrocarbons would interact with
nonpolar air, polar heads with
water at surface.

7.75   Phospholipids are amphipathic molecules. They act to emulsify the oil and the water in the mayonnaise, forming a thick mixture.

Copyright © 2014 Pearson Education, Inc.

7.77   a. Honey

b. Hydrogen bonding, dipole–dipole attractions, and London forces. A glucose in ring form can make many more hydrogen bonds to water.

c. Because there are so many more hydrogen bonds possible between sugar molecules in honey, this makes the honey more viscous than water.

d. By adding soap to the water, the hydrogen bonding network between the water molecules is disrupted, and the paper clip can no longer float on the surface.

Copyright © 2014 Pearson Education, Inc.

Solution Chemistry: How Sweet Is Your Tea?

8.1 Solutions Are Mixtures

Learning Objectives

Upon completion of this material, a student should be able to do the following:

A. Define the following key terms:

solution	**aqueous solution**
solute	**colloid**
solvent	**suspension**

B. Distinguish solute and solvent.

C. Identify solutions, colloids, and suspensions.

8.2 Formation of Solutions

Learning Objectives

Upon completion of this material, a student should be able to do the following:

A. Define the following key terms:

solubility	**saturated solution**
solvation	**equilibrium**
hydration	**Henry's law**
unsaturated solution	

B. Define saturated and dilute solutions.

C. Predict the effect of temperature on the solubility of a solute.

D. Predict the effect of pressure on the solubility of a gas in a liquid.

8.3 Chemical Equations for Solution Formation

Learning Objectives

Upon completion of this material, a student should be able to do the following:

A. Define the following key terms:

electrolyte	**weak electrolyte**
strong electrolyte	**law of conservation of mass**
ionize	**equivalent (Eq)**
nonelectrolyte	

B. Write chemical equations for hydration of electrolytes, nonelectrolytes, and weak electrolytes.

C. Calculate the number of milliequivalents present for an ionic compound that fully dissociates in solution.

Copyright © 2014 Pearson Education, Inc.

D. Convert from mEq to moles.

8.4   Concentrations

Learning Objectives
Upon completion of this material, a student should be able to do the following:
   A. Define the following key terms:
       concentration   **molarity**
   B. Express concentration in molarity units.
   C. Express concentration in percent units.
   D. Express concentration in parts per million and parts per billion.

8.5   Dilution

Learning Objectives
Upon completion of this material, a student should be able to do the following:
   A. Calculate concentrations or determine volumes using the dilution equation.

8.6 Osmosis and Diffusion

Learning Objectives
Upon completion of this material, a student should be able to do the following:
   A. Define the following key terms:

semipermeable membrane	**hypertonic solution**
isotonic solution	**crenation**
hypotonic solution	**physiological solution**
osmosis	**diffusion**
osmotic pressure	**dialysis**

   B. Predict the direction of osmosis or diffusion given the concentration on both sides of a semipermeable membrane.

8.7 Transport Across Cell Membranes

Learning Objectives
Upon completion of this material, a student should be able to do the following:
   A. Define the following key terms:

passive diffusion	**active transport**
facilitated transport	

   B. Characterize three forms of transport across a cell membrane.

Copyright © 2014 Pearson Education, Inc.

Practice Test for Chapter 8

1. In which of the following would ethanol be classified the solvent?
   A. 35 mL of ethanol is mixed with 65 mL of methanol.
   B. 100 mL of ethanol is mixed with enough water to form 115 mL of solution.
   C. 35 g of ethanol is mixed with enough isopropyl alcohol to produce 150 g of solution.
   D. Ethanol is not the solvent in any of the solutions described.

2. Honey is mixed in hot tea. The resulting combination of honey and hot tea would be classified as:
   A. suspension.
   B. colloid.
   C. solution.
   D. solvent.

3. Blood is a _____ of plasma and blood cells that can be separated by centrifugation. Fill the blank with one of the choices.
   A. suspension
   B. colloid
   C. solution
   D. None of the above correctly classifies blood.

4. What happens to the solubility of oxygen in blood when a person drives from Kona, Hawaii at sea level to the top of Mauna Kea at almost 15,000 ft elevation?
   A. decreases
   B. increases
   C. remains constant
   D. cannot predict from the information given

5. Consider two identical bottles of soft drink. One is stored in the refrigerator and one is stored at room temperature. When opened, which one will release more gas **from the liquid**?

   A. the one stored at room temperature
   B. the one stored in the refrigerator
   C. Neither, they will release the same amount of gas regardless of temperature.
   D. Cannot predict based on the given information.

Copyright © 2014 Pearson Education, Inc.

6. The following picture was taken after table salt was added to water after the mixture was thoroughly stirred for 5 minutes. The pictured mixture probably represents a(n):

A. heterogeneous mixture.
B. saturated solution.
C. equilibrium state.
D. all of the above

7. Consider the following image of containers of the same carbonated beverage.

In which container is the pressure of CO_2 likely to be the smallest?
A. 2 L bottle
B. beaker
C. 12 oz can
D. impossible to predict

8. For **both** solid and gaseous solutes, which of the following is true?
A. Solubility increases with increasing temperature.
B. Solubility decreases with increasing temperature.
C. Solubility increases with increasing pressure.
D. Solubility decreases with increasing pressure.
E. None of the above is true for both solid and gaseous solutes.

9. Which of the following is most likely to be a weak electrolyte?
A. $C_6H_{12}O_6(s)$
B. $NH_4NO_3(s)$
C. $CH_3COOH(l)$
D. $CH_3OH(l)$

Copyright © 2014 Pearson Education, Inc.

10. When HF(l) dissolves in water, the following reaction occurs. HF is classified as:

$$HF(l) \rightleftharpoons H^+(aq) + F^-(aq)$$
$$H_2O$$

    A. weak electrolyte.
    B. nonelectrolyte.
    C. strong electrolyte.
    D. The reaction alone cannot be used to predict.

11. The following reaction indicates:

$$MgCl_2(s) \xrightarrow[H_2O]{} Mg^{2+}(aq) + 2\,Cl^-(aq)$$

    A. that the solubility of the solute will decrease as the temperature is increased.
    B. that the solute is a weak electrolyte.
    C. that the reaction is a hydration.
    D. all of the above

12. The balanced equation for the hydration of $AlCl_3$ would contain how many total ions or molecules in the products?
    A. 1
    B. 2
    C. 3
    D. 4
    E. 5

13. The products of the correctly balanced equation for the hydration of $BaCl_2$ would be:
    A. $Ba^{2+}(aq) + Cl_2^{2-}(aq)$.
    B. $Ba^{2+}(aq) + 2Cl^-(aq)$.
    C. $BaCl_2(aq)$.
    D. $Ba^{1+}(aq) + 2Cl^-(aq)$.

14. Which of the following concentration units is used to represent the amount of electrolytes in body fluid?
    A. M
    B. %(m/v)
    C. %(m/m)
    D. mEq/L

Copyright © 2014 Pearson Education, Inc.

15. Which of the following conversion factors could be used to represent the relationship between equivalents and moles of PO_4^{3-}?

A. $\dfrac{1\,Eq\,PO_4^{3-}}{1\,mol\,PO_4^{3-}}$

B. $\dfrac{1\,Eq\,PO_4^{3-}}{3\,mol\,PO_4^{3-}}$

C. $\dfrac{3\,Eq\,PO_4^{3-}}{1\,mol\,PO_4^{3-}}$

D. $\dfrac{3\,Eq\,PO_4^{3-}}{3\,mol\,PO_4^{3-}}$

16. How many equivalents of CO_3^{2-} are present in a solution that contains 0.750 moles of CO_3^{2-}?
    A. 1.50 Eq
    B. 0.750 Eq
    C. 0.250 Eq
    D. 2.25 Eq

17. How many mmol of Ca^{2+} are present in a solution that contains 2.30 mEq of CO_3^{2-}?
    A. 2.30 mmol
    B. 4.60 mmol
    C. 1.15 mmol
    D. 6.90 mmol

18. A Ringer's solution for intravenous fluid replacement typically has a concentration of 77.5 mEq Cl^- per 0.500 L of solution. If a patient receives 0.750 L of Ringer's solution, how many equivalents of Cl^- were given?

    A. 116 Eq
    B. 0.116 Eq
    C. 51. 7 Eq
    D. 0.0775 Eq

19. What is the molarity of a solution prepared by dissolving 2.50 moles of NaCl in enough water to produce 3.00 L solution?
    A. 0.833 M
    B. 2.50 M
    C. 7.50 M
    D. 1.20 M

20. How many moles of KCl are present in 574 mL of a 0.660 M KCl solution?
    A. 0.660 mol
    B. 379 mol
    C. 1.15 mol
    D. 0.379 mol

Copyright © 2014 Pearson Education, Inc.

21. Which of the following concentration scales would be the most useful for expressing the concentration of very dilute solutions?

    A. M
    B. % (m/m)
    C. ppm
    D. ppb
    E. either C or D

22. What is the molarity of a solution prepared by dissolving 50.0 g of KCl in enough water to yield 1.50 L of solution?

    A. 33.3 M
    B. 0.0300 M
    C. 0.447 M
    D. 0.671 M

23. Calculate the concentration in %(m/m) for a solution prepared by dissolving 58.5 g of Na_2SO_4 in 200.0 g of distilled water.

    A. 22.6%
    B. 29.2%
    C. 3.41%
    D. 11.7%

24. 0.125 L of a 55.0 mg/mL hydrocortisone solution is needed. What volume of a 210.0 mg/ml solution should be diluted to obtain this solution?

    A. 32.7 mL
    B. 477 mL
    C. 0.0327 mL
    D. 6.88 mL
    E. 26.2 mL

25. How many grams of glucose would be needed to prepare 2.00 L of a 5.10% (m/v) solution?
    A. 10.2 g
    B. 102 g
    C. 2.55 g
    D. 25.5 g

26. How many liters of a 5.00% (m/v) solution of NaCl can be prepared from 575 mL of a stock 12.5% (m/v) solution?
    A. 230 L
    B. 1440 L
    C. 0.230 L
    D. 1.44 L

Copyright © 2014 Pearson Education, Inc.

27. In which of the following processes is the direction of ion or molecule movement opposite to that which will equalize concentrations?
   A. osmosis
   B. diffusion
   C. active transport
   D. passive diffusion
   E. none of these

28. Consider the following image, which depicts a cell in a solution.

   Based on the effect on the cell, the solution would be classified as:
   A. hypertonic and the water is flowing out of the cell.
   B. hypotonic and the water is flowing into the cell.
   C. hypertonic and the water is flowing into the cell.
   D. hypotonic and the water is flowing out of the cell.
   E. isotonic and there is no net water flow.

29. Consider the diagram showing methods of transport across a cell membrane.

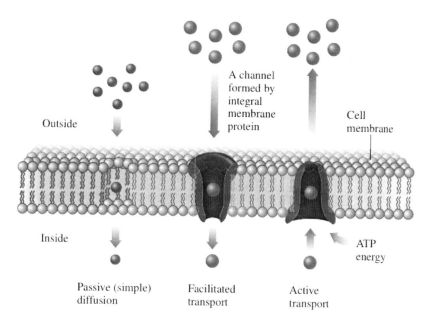

   A Na^+ ion is to be transported across the membrane with an input of energy. Which method of transport will be used?

Copyright © 2014 Pearson Education, Inc.

A. passive diffusion
B. facilitated transport
C. active transport
D. Not enough information is given to determine the method.

30. A red blood cell that is isotonic with 0.90% (m/v) NaCl and 5% (m/v) glucose will undergo:
    A. lysis if placed in distilled water.
    B. crenation if placed in a 1.90% (m/v) NaCl and 5% (m/v) glucose solution.
    C. no change if placed in a 0.90% (m/v) NaCl and 5% (m/v) glucose solution.
    D. all of the above

Copyright © 2014 Pearson Education, Inc.

Answers

1. B   2. C   3. A   4. A   5. B   6. D   7. B   8. E   9. C   10. A
11. C   12. D   13. B   14. D   15. C   16. A   17. C   18. B   19. A   20. D
21. E   22. C   23. A   24. A   25. B   26. D   27. C   28. A   29. C   30. D

Copyright © 2014 Pearson Education, Inc.

Chapter 8 – Solutions to Odd-Numbered Problems

Practice Problems

8.1   a. solute, oxygen; solvent, nitrogen
b. solute, zinc; solvent, copper
c. solute, blue food coloring; solvent, ethanol

8.3   a. colloid   b. solution   c. solution

8.5   a. no, not saturated   b. yes, saturated

8.7   a. increase   b. decrease   c. increase

8.9   a. Decreasing the temperature increases the solubility of the gas in the soda, so more of the gas stays in the solution, and capping the bottle increases the CO_2 pressure over the solution and increases the solubility of CO_2.
b. If the lid is on tight, no gas can escape, and gas that escapes from the solution will build up a pressure above the solution. At some point, an equilibrium is reached in which no more gas will escape from the solution.
c. The solubilities of most solid solutes decrease with lower temperature, so less sugar will dissolve in the iced tea.

8.11   a. fully dissociate   b. partially dissociate c. not dissociate

8.13   a.

$$KF(s) \xrightarrow{H_2O} K^+(aq) + F^-(aq)$$

b.

$$HCN(g) \underset{H_2O}{\rightleftharpoons} H^+(aq) + CN^-(aq)$$

c.

$$C_6H_{12}O_6(s) \xrightarrow{H_2O} C_6H_{12}O_6(aq)$$

8.15   a.

$$CaCl_2(s) \xrightarrow{H_2O} Ca^{2+}(aq) + 2Cl^-(aq)$$

b.

$$NaOH(s) \xrightarrow{H_2O} Na^+(aq) + OH^-(aq)$$

c.

$$KBr(s) \xrightarrow{H_2O} K^+(aq) + Br^-(aq)$$

d.

$$Fe(NO_3)_3(s) \xrightarrow{H_2O} Fe^{3+}(aq) + 3NO_3^-(aq)$$

8.17   4.25 Eq

8.19   0.0770 mole Na^+

8.21   1.25 mmoles $Ca2^+$/L

8.23   17 M

8.25   0.43 M

8.27   35.7 g of KBr

8.29   0.5% (m/v)

8.31   a. 5.00%   b. 6.3%

8.33   0.117 g insulin

8.35   5 ppm, 5000 ppb

Copyright © 2014 Pearson Education, Inc.

8.37 6 L

8.39 3% (m/v)

8.41 Add 62.5 mL of the 0.90% NaCl stock solution to enough distilled water for a total volume of 250 mL of solution.

8.43 a. leave   b. crenate   c. hypertonic

8.45 hypertonic, isotonic

8.47 a. hypertonic   b. hypotonic   c. hypertonic   d. isotonic

8.49 a. passive diffusion
    b. facilitated transport
    c. active transport
    d. facilitated transport

Additional Problems

8.51 a. colloid   b. solution   c. colloid

8.53 a. decrease   b. increase   c. decrease

8.55 a. fully dissociate   b. partially dissociate   c. not dissociate

8.57 a.

$$NaI(s) \xrightarrow{H_2O} Na^+(aq) + I^-(aq)$$

    b.

$$HCOOH(l) \underset{H_2O}{\rightleftharpoons} HCOO^-(aq) + H^+(aq)$$

    c.

$$C_6H_{12}O_6(s) \xrightarrow{H_2O} C_6H_{12}O_6(aq)$$

8.59 2.68 Eq Ca^{2+}

8.61 7 mEq Ca^{2+}

8.63 12.5 mmole/L

8.65 1.87 M

8.67 0.80 M

8.69 90. mL of ethanol

8.71 25 g of dextrose

8.73 70–180 ppm urea nitrogen

8.75 a. 1.0 M HNO_3
    b. 0.75 M KOH

8.77 a. 0.90 L
    b. 1.25 L

8.79 Add 50 mL of the stock solution (18% (m/v) to enough water to make 1.0 L.

8.81 a. hypertonic
    b. hypotonic
    c. isotonic

8.83 The cells will attempt to dilute the higher than normal concentration of Na^+ in the tissues by moving more water into the tissues causing fluid retention.

8.85 If osmosis is water moving from a lower concentrated solution to a higher concentrated solution, balancing out the concentrations, the reverse would require water moving in the opposite direction. Less pure water (higher concentration of solutes) is forced (requires energy) through a filter (semipermeable membrane), removing more impurities, making drinkable water.

Copyright © 2014 Pearson Education, Inc.

8.87 a. no   b. no   c. yes

Challenge Problems

8.89 According to Henry's law, more nitrogen would dissolve in the bloodstream at lower depths (higher pressure) than at the surface (lower pressure). When a diver ascends to the surface quickly, the pressure of the air in the tank (and therefore the air in the lungs) lessens more rapidly than a person can expel the nitrogen. Unable to stay dissolved in the bloodstream, nitrogen gas bubbles begin forming in the bloodstream, causing the condition.

8.91 Water will move from solution A to solution B because solution B is a higher concentrated solution (more solute/solution). It does not matter that the albumin particles are huge; osmotic flow depends on the concentration, not the size of the particles. Since solution B has a higher solute concentration, it exerts a higher osmotic pressure.

Copyright © 2014 Pearson Education, Inc.

Chapter

9

Acids, Bases, and Buffers in the Body

9.1 Acids and Bases–Definitions

Learning Objectives

Upon completion of this material, a student should be able to do the following:
  A. Define the following key terms:

  acid   **base**
  hydronium ion, H_3O^+

  B. Describe an acid and a base using the Arrhenius or Brønsted-Lowry definition.
  C. Describe the physical characteristics of an acid and a base.

9.2 Strong Acids and Bases

Learning Objectives

Upon completion of this material, a student should be able to do the following:
  A. Define the following key terms:

  strong acid   **weak base**
  weak acid   **salt**
  strong base   **neutralization**

  B. Name the six strong acids.
  C. Compare a strong acid to a weak acid.
  D. Write and balance a neutralization reaction.

9.3 Chemical Equilibrium

Learning Objectives

Upon completion of this material, a student should be able to do the following:
  A. Define the following key terms:

  chemical equilibrium   **Le Châtelier's principle**
  equilibrium constant

  B. Define chemical equilibrium.
  C. Write an equilibrium expression for K.
  D. Apply Le Châtelier's principle to chemical equilibrium.

9.4 Weak Acids and Bases

Learning Objectives

Upon completion of this material, a student should be able to do the following:
  A. Define the following key terms:

  acid dissociation constant, K_a   **conjugate acid**

Copyright © 2014 Pearson Education, Inc.

conjugate base   **conjugate acid-base pair**

B. Apply the principles of chemical equilibrium to weak acids and bases.
C. Determine strengths of weak acids based on their pK_a values.
D. Identify conjugate acid-base pairs.
E. Complete a chemical equation for a conjugate acid-base in water.

9.5   pH and the pH Scale

Learning Objectives
Upon completion of this material, a student should be able to do the following:
A. Define the following key terms:

autoionization of water   **basic**
acidic   **pH**
neutral

B. Determine if a solution is acidic, basic, or neutral if given its pH.
C. Calculate the pH if given the $[H_3O^+]$.
D. Calculate the $[H_3O^+]$ if given the pH.

9.6 pK_a

Learning Objectives
Upon completion of this material, a student should be able to do the following:
A. Define the following key terms.

pK_a

B. Predict the strength of a weak acid from its pK_a value.

9.7   Amino Acids: Common Biological Weak Acids

Learning Objectives
Upon completion of this material, a student should be able to do the following:
A. Define the following key terms:

amino acids   **zwitterion**
isoelectric point (pI)

B. Define isoelectric point.
C. Predict the charge of an amino acid below, at, and above the pI value.

9.8   Buffers and Blood: The Bicarbonate Buffer System

Learning Objectives
Upon completion of this material, a student should be able to do the following:
A. Define the following key terms:

buffer   **hyperventilation**
homeostasis   **respiratory alkalosis**
hypoventilation   **metabolic acidosis**
respiratory acidosis   **metabolic alkalosis**

B. Describe the properties of a buffer.
C. Predict the direction the bicarbonate buffer equilibrium will shift with changes in ventilation rate.

Copyright © 2014 Pearson Education, Inc.

Practice Test for Chapter 9

1. Which of the following describes a base?
   A. slippery feel
   B. sour taste
   C. corrodes metal
   D. donates a proton

2. An acid can be described as producing which of the following?
   A. proton
   B. H^+ ion
   C. H_3O^+ ion
   D. all of these

3. For the following reaction:
$$NH_3(aq) + H_3O^+(aq) \rightarrow H_2O(l) + NH_4^+(aq)$$
   which of the following is a correct statement?
   A. NH_3 is the acid and H_3O^+ is the base.
   B. NH_3 is the base and H_3O^+ is the acid.
   C. H_2O and H_3O^+ are both acids.
   D. NH_3 and NH_4^+ are both bases.

4. In which of the following reactions does water act as a proton donor?
   A. $HBr(aq) + H_2O(l) \rightarrow H_3O^+(aq) + Br^-(aq)$

   B. $H_2O(l) + CN^-(aq) \rightarrow HCN(aq) + OH^-(aq)$

   C. $HCl(g) + H_2O(l) \rightarrow H_3O^+(aq) + Cl^-(aq)$

   D. none of the above

5. Which of the following would be classified as a strong acid?
   A. HF
   B. HNO_2
   C. HCl
   D. H_3PO_4
   E. All of the above are strong acids.

6. Which of the following compounds will completely ionize in water?
   A. NH_3
   B. H_2CO_3
   C. CH_3COOH
   D. KOH

Copyright © 2014 Pearson Education, Inc.

7. When the equation is balanced, the products of the following reaction are:

$$H_2SO_4 + Ca(OH)_2 \rightarrow$$

A. $CaSO_4 + 2\,H_2O$
B. $CaSO_4 + H_2O$
C. $SO_4(OH)_2 + CaH_2$
D. $2SO_4OH + 2CaH$

8. What is the equilibrium constant expression for the following reaction?

$$CaCO_3(s) + 2\,HI(aq) \rightleftharpoons CaI_2(aq) + CO_2(g) + H_2O(l)$$

A. $K = \dfrac{[CaI_2][CO_2][H_2O]}{[CaCO_3][HI]}$

B. $K = \dfrac{[CaI_2][CO_2][H_2O]}{[CaCO_3][HI]^2}$

C. $K = \dfrac{[CaI_2][CO_2]}{[HI]}$

D. $K = \dfrac{[CaI_2][CO_2]}{[HI]^2}$

9. At a certain temperature, K for the following reaction is 54.

$$H_2(g) + I_2(g) \rightleftharpoons 2HI(g)$$

This means at equilibrium
A. the products are present in greater amounts than the reactants.
B. the reactants are present in greater amounts than the products.
C. the products and reactants are present in equal amounts.
D. the amounts of reactants and products will vary depending on the starting amount of reactant.

10. Consider the following reaction.

$$2\,SO_2(g) + O_2(g) \rightleftharpoons 2\,SO_3(g) + heat$$

Which of the following will increase the amount of SO_3?
A. increasing the temperature
B. removing O_2
C. adding SO_2
D. All increase the amount of SO_3.

11. Which of the following changes will increase the amount of product at equilibrium for an endothermic reaction?
A. remove the product as it forms
B. increase the concentration of a reactant
C. increase the temperature of the reaction
D. A and B
E. All of the above will increase the amount of product.

Copyright © 2014 Pearson Education, Inc.

12. Rank the following acids in order of increasing strength, weakest to strongest (left to right).

$$HCN \qquad HCl \qquad HNO_2 \qquad NH_4^+$$

TABLE 9.5 K_a Values for Substances Acting as Weak Acids (25 °C)

Name	Formula	K_a
Hydrogen sulfate ion	HSO_4^-	$1.0 * 10^{-2}$
Phosphoric acid	H_3PO_4	$7.5 * 10^{-3}$
Hydrofluoric acid	HF	$6.5 * 10^{-4}$
Nitrous acid	HNO_2	$4.5 * 10^{-4}$
Formic acid	$HCOOH$	$1.8 * 10^{-4}$
Acetic acid	CH_3COOH	$1.75 * 10^{-5}$
Carbonic acid	H_2CO_3	$4.5 * 10^{-7}$
Water	H_2O	$1.0 * 10^{-7}$
Dihydrogen phosphate ion	$H_2PO_4^-$	$6.6 * 10^{-8}$
Ammonium ion	NH_4^+	$6.3 * 10^{-10}$
Hydrocyanic acid	HCN	$6.2 * 10^{-10}$
Bicarbonate ion	HCO_3^-	$4.8 * 10^{-11}$
Hydrogen phosphate ion	HPO_4^{2-}	$1.0 * 10^{-12}$

Increasing acid strength

A. $HCN < HNO_2 < NH_4^+ < HCl$
B. $HCN < NH_4^+ < HNO_2 < HCl$
C. $HCl < HNO_2 < NH_4^+ < HCN$
D. $HCl \approx HNO_2 < NH_4^+ < HCN$

13. What is the conjugate base of HCN in the following reaction?
$$H_2O(l) + CN^-(aq) \rightleftharpoons HCN(aq) + OH^-(aq)$$

A. H_2CN^+
B. OH^-
C. H_2O
D. CN^-

14. Which of the following is true of HCO_3^-?
A. CO_3^{2-} is the conjugate base.
B. H_2CO_3 is the conjugate acid.
C. H_2O is the conjugate base.
D. H_3O^+ is the conjugate acid.
E. A and B

Copyright © 2014 Pearson Education, Inc.

15. The products of the following reaction would be:

$$F^-(aq) + H_2O(l) \rightleftharpoons$$

   A. $HF(aq) + H_3O^+(aq)$
   B. $H_2F^+(aq) + OH^-(aq)$
   C. $HF(aq) + OH^-(aq)$
   D. $H_2F^+(aq) + H_3O^+(aq)$

16. What is the log of 1.75×10^{-3} and the inverse (INV or 10^x) log of -4.55, respectively?
   A. 0.243, 3.54×10^4
   B. -2.76, 2.82×10^{-5}
   C. 2.76, 2.82×10^{-5}
   D. 3.24, 5.5×10^{-4}
   E. -2.76, It is not possible to take the inverse log of a negative number.

17. A sample of seawater has a pH of 8.23. This solution is:
   A. basic.
   B. acidic.
   C. neutral.

18. A grapefruit has a $[H_3O^+]$ of 3.35×10^{-3}. This solution is:
   A. basic.
   B. acidic.
   C. neutral.

19. What is the pH of solution of coffee with $[H_3O^+]$ of 6.3×10^{-5}?
   A. 9.67
   B. -4.20
   C. 1.30
   D. 4.20

20. What is the $[H_3O^+]$ of a baking soda solution with a pH of 9.22?
   A. 2.2×10^{-9}
   B. 9.65×10^{-1}
   C. 6.03×10^{-10}
   D. 1.01×10^4

21. As the pK_a of an acid increases, the strength of the acid:
   A. increases.
   B. decreases.
   C. remains constant.

22. The pK_a of HF is 3.19. At a pH of 5.25,
   A. $[HF] = [F^-]$.
   B. $[HF] > [F^-]$.
   C. $[HF] < [F^-]$.
   D. Relative concentrations cannot be determined from the given information.

Copyright © 2014 Pearson Education, Inc.

23. The pK_a of HCN is 9.12. If [HCN] = [CN⁻], the pH of the solution must be:
A. 9.12.
B. > 9.12.
C. < 9.12.
D. between 7.00 and 9.12.

24. Which of the following has the strongest acid of the pair listed on the left?

TABLE 9.7 pK_a and K_a Values for Substances Acting as Weak Acids (25 °C)

Name	Formula	pK_a	K_a
Hydrogen sulfate ion	HSO_4^-	2.00	1.0×10^{-2}
Phosphoric acid	H_3PO_4	2.12	7.5×10^{-3}
Hydrofluoric acid	HF	3.19	6.5×10^{-4}
Nitrous acid	HNO_2	3.35	4.5×10^{-4}
Formic acid	HCOOH	3.74	1.8×10^{-4}
Acetic acid	CH_3COOH	4.76	1.75×10^{-5}
Carbonic acid	H_2CO_3	6.35	4.5×10^{-7}
Water	H_2O	7.00	1.0×10^{-7}
Dihydrogen phosphate ion	$H_2PO_4^-$	7.18	$6.6 * 10^{-8}$
Ammonium ion	NH_4^+	9.20	6.3×10^{-10}
Hydrocyanic acid	HCN	9.21	6.2×10^{-10}
Bicarbonate ion	HCO_3^-	10.32	4.8×10^{-11}
Hydrogen phosphate ion	HPO_4^{2-}	12.00	1.0×10^{-12}

Increasing acid strength

A. HCO_3^- and H_2CO_3
B. HPO_4^{2-} and $H_2PO_4^-$
C. $H_2PO_4^-$ and H_3PO_4
D. All pairs have the stronger acid listed on the right.

Copyright © 2014 Pearson Education, Inc.

25. Using the pK_a values in the following table, predict the products of the following reaction.

$$HCO_3^-(aq) + H_2PO_4^{2-}(aq) \rightleftharpoons$$

TABLE 9.7 pK_a and K_a Values for Substances Acting as Weak Acids (25 °C)

Name	Formula	pK_a	K_a
Hydrogen sulfate ion	HSO_4^-	2.00	1.0×10^{-2}
Phosphoric acid	H_3PO_4	2.12	7.5×10^{-3}
Hydrofluoric acid	HF	3.19	6.5×10^{-4}
Nitrous acid	HNO_2	3.35	4.5×10^{-4}
Formic acid	$HCOOH$	3.74	1.8×10^{-4}
Acetic acid	CH_3COOH	4.76	1.75×10^{-5}
Carbonic acid	H_2CO_3	6.35	4.5×10^{-7}
Water	H_2O	7.00	1.0×10^{-7}
Dihydrogen phosphate ion	$H_2PO_4^-$	7.18	$6.6 * 10^{-8}$
Ammonium ion	NH_4^+	9.20	6.3×10^{-10}
Hydrocyanic acid	HCN	9.21	6.2×10^{-10}
Bicarbonate ion	HCO_3^-	10.32	4.8×10^{-11}
Hydrogen phosphate ion	HPO_4^{2-}	12.00	1.0×10^{-12}

Increasing acid strength

A. $CO_3^{2-}(aq) + H_3PO_4(aq)$
B. $H_2CO_3(aq) + HPO_4^{2-}(aq)$
C. $H_2CO_3(aq) + PO_4^{3-}(aq)$
D. The products cannot be predicted.

26. For an amino acid, the pK_a for the carboxylic acid group is 4.50 and pK_a for the protonated amine group is 11.00. What is the pI for this amino acid?
A. 4.50
B. 11.00
C. 15.50
D. 6.50
E. 7.75

27. Which of the following represents the zwitterion of the amino acid valine?
A.

Copyright © 2014 Pearson Education, Inc.

B.

$$H_2N-\underset{\underset{H}{|}}{\overset{\overset{\displaystyle CH(CH_3)_2}{|}}{C}}-\underset{}{\overset{\overset{\displaystyle O}{\|}}{C}}-O^-$$

C.

$$H_3\overset{+}{N}-\underset{\underset{H}{|}}{\overset{\overset{\displaystyle CH(CH_3)_2}{|}}{C}}-\underset{}{\overset{\overset{\displaystyle O}{\|}}{C}}-O^-$$

D. None of these represents the zwitterion.

28. Under what conditions would the following reaction occur?

$$H_3\overset{+}{N}-\underset{\underset{H}{|}}{\overset{\overset{\displaystyle CH_3}{|}}{C}}-\underset{}{\overset{\overset{\displaystyle O}{\|}}{C}}-OH \longrightarrow H_2N-\underset{\underset{H}{|}}{\overset{\overset{\displaystyle CH_3}{|}}{C}}-\underset{}{\overset{\overset{\displaystyle O}{\|}}{C}}-O^-$$

    A. addition of an acid
    B. decrease in the pH
    C. increase in the hydronium ion concentration
    D. addition of a proton acceptor

29. Hyperventilation:
    A. is also known as respiratory alkalosis.
    B. results in decreased amounts of CO_2 in the blood.
    C. increases the pH of the blood.
    D. upsets the bicarbonate balance of the blood.
    E. All of the above are characteristic of hyperventilation.

30. The blood buffer system can be represented as:

$$CO_2(g) + H_2O(l) \rightleftharpoons H_3O^+(aq) + HCO_3^-(aq)$$

Vomiting can cause a loss of the hydronium ion from blood. This causes:
    A. metabolic alkalosis.
    B. a shift in the reaction toward the formation of more CO_2.
    C. a decrease in blood pH.
    D. an imbalance that can be treated with bicarbonate solution.
    E. all of the above

Copyright © 2014 Pearson Education, Inc.

Answers

1. A   2. D   3. B   4. B   5. C   6. D   7. A   8. D   9. A   10. C

11. E   12. B   13. D   14. E   15. C   16. B   17. A   18. B   19. D   20. C

21. B   22. C   23. A   24. D   25. B   26. E   27. C   28. D   29. E   30. A

Copyright © 2014 Pearson Education, Inc.

Chapter 9 – Solutions to Odd-Numbered Problems

Practice Problems

9.1 a. acid   b. base   c. acid   d. base

9.3 Most hydrogen atoms contain one proton, one electron, and no neutrons. Therefore, a positively charged hydrogen ion (H^+) is simply one proton.

9.5 a. HI is the acid (proton donor) and H_2O is the base (proton acceptor).
b. H_2O is the acid (proton donor) and F^- is the base (proton acceptor).

9.7 The strong acids are (a.) H_2SO_4, (b.) HCl, and (d.) HNO_3.

9.9 a.

$$HNO_3(aq) + LiOH(s) \rightarrow H_2O(l) + LiNO_3(aq)$$

b.

$$H_2SO_4(aq) + Ca(OH)_2(s) \rightarrow 2H_2O(l) + CaSO_4(aq)$$

9.11 a.

$$3HBr(aq) + Al(OH)_3(s) \rightarrow 3H_2O(l) + AlBr_3(aq)$$

b.

$$2HI(aq) + CaCO_3(s) \rightarrow H_2O(l) + CaI_2(aq) + CO_2(g)$$

9.13 Reversible reactions are reactions that can proceed in both the forward and reverse directions.

9.15 a. $K = \dfrac{[CO_2][H_2]}{[CO][H_2O]}$ b. $K = \dfrac{[CH_3COO^-][H_3O^+]}{[CH_3COOH]}$

9.17 a. reactants
b. products
c. both reactants and products present in equal amounts

9.19 a. products are favored.   b. reactants are favored.
c. products are favored.   d. reactants are favored.

9.21 a. reactants are favored.   b. products are favored.
c. products are favored.   d. reactants are favored.

9.23 The escape of CO_2 gas removes CO_2; equilibrium favors the formation of the products. The soda no longer bubbles.

9.25 a. $H_2PO_4^-$ b. HF   c. HBr

9.27 a. acid HSO_4^- conjugate base SO_4^{2-}
base H_2O conjugate acid H_3O^+
b. acid NH_4^+ conjugate base NH_3
base H_2O conjugate acid H_3O^+
c. acid HCN; conjugate base CN^-
base NO_2^- conjugate acid HNO_2

9.29 a. F^- fluoride ion   b. OH^- hydroxide ion
c. HCO_3^- bicarbonate ion   d. SO_4^{2-} sulfate ion

9.31 a. HCO_3^- bicarbonate ion   b. H_3O^+ hydronium ion
c. H_3PO_4 phosphoric acid   d. HBr, hydrobromic acid

Copyright © 2014 Pearson Education, Inc.

9.33. a.

$$HA(aq) + H_2O(l) \rightleftharpoons A^-(aq) + H_3O^+(aq)$$

Acid   Base   Conjugate   Conjugate
base   acid

b.

$$H_2PO_4^-(aq) + H_2O(l) \rightleftharpoons HPO_4^{2-}(aq) + H_3O^+(aq)$$

Acid   Base   Conjugate   Conjugate
base   acid

c.

$$NH_3(aq) + H_2O(l) \rightleftharpoons NH_4^+(aq) + OH^-(aq)$$

Base   Acid   Conjugate   Conjugate
acid   base

9.35   a. basic   b. acidic   c. basic   d. acidic
9.37   a. basic   b. acidic   c. basic   d. neutral
9.39   a. 7.92   b. 2.15   c. 10.33   d. 7.00

9.41   a. 7.9×10^{-13} M   b. 3×10^{-6} M   c. 1.0×10^{-2} M   d. 5×10^{-9} M

9.43   a. $H_2PO_4^-$ b. H_2SO_4 c. formic acid   d. ammonium ion

9.45   a.   b.   c.

9.47   a. Hyperventilation will lower the CO_2 level in the blood, which decreases the H_3O^+ and increases the blood pH.
b. The equilibrium will shift to the left.
c. This condition is called respiratory alkalosis.
d. Breathing into a bag will increase the CO_2 level, which increases H_3O^+ (by shifting the equilibrium to the right), and lowers the blood pH.
e. The equilibrium will shift to the right.

Additional Problems

9.49   a. strong   b. weak   c. strong   d. weak
9.51   Both strong and weak acids produce H_3O^+ in water. Weak acids are only slightly ionized, whereas a strong acid exists only as ions in solution (fully ionizes).
9.53   a.   b. HI predominates.

$$K = \frac{[H_2]\,[I_2]}{[HI]^2}$$

Copyright © 2014 Pearson Education, Inc.

9.55   a. exothermic

$$K = \frac{[CH_3OH]}{[CO][H_2]^2}$$
b.

c. It will shift to the right.
d. It will shift to the left.
e. It will shift to the right.

9.57 a. NO_2^- b. CH_3NH_2 c. $HCOO^-$

9.59 a. $CH_3CH(OH)COOH$ b. $CH_3NH_3^+$ c. HPO_4^{2-}

9.61   a. 7.60   b. 1.15   c. 10.4

9.63   a. 3.0×10^{-4} M   b. 1×10^{-5} M   c. 5.6×10^{-10} M

9.65   a. A buffer system keeps the pH constant.
b. The conjugate base CH_3COO^- from the salt $NaCH_3COO$ is needed to neutralize any added acid.
c. Added H_3O^+ reacts with CH_3COO^-.
d. Added OH^- reacts with CH_3COOH.

9.67   a. metabolic acidosis
b. below normal

c. administer bicarbonate (HCO_3^-)

9.69   a.   b.

pH = 10.0

pH = 3.0

Challenge Problems

9.71

9.73   Number of moles of NaOH = number of moles of acetic acid

$$\frac{0.500 \text{ moles NaOH}}{1 \cancel{L}} \times \frac{1 \cancel{L}}{1000 \cancel{mL}} \times 16.5 \cancel{mL} = 0.00825 \text{ moles NaOH}$$

$$\frac{0.00825 \text{ moles acetic acid}}{10.0 \cancel{mL}} \times \frac{1000 \cancel{mL}}{1 \cancel{L}} = 0.825 \text{ M acetic acid}$$

Copyright © 2014 Pearson Education, Inc.

Proteins — Workers of the Cell

10.1   Amino Acids—A Second Look

Learning Objectives

Upon completion of this material, a student should be able to do the following:
   A. Define the following key terms:

   alpha (α) carbon   **nonpolar amino acids**
   alpha (α) amino group   **polar amino acids**
   alpha (α) carboxylate group
   B. Draw the general structure of an amino acid.
   C. Identify amino acids based on their polarity.

10.2 Protein Formation

Learning Objectives

Upon completion of this material, a student should be able to do the following:
   A. Define the following key terms:

   amide   **C-terminus**
   peptide bond   **polypeptide**
   dipeptide   **protein**
   N-terminus
   B. Predict the products of a biological condensation or hydrolysis reaction.
   C. Form a peptide bond between amino acids.

10.3 The Three-Dimensional Structure of Proteins

Learning Objectives

Upon completion of this material, a student should be able to do the following:
   A. Define the following key terms:

   primary (1°) structure   **hydrophobic effect**
   protein backbone   **polar interactions**
   secondary (2°) structure   **salt bridges (ionic attractions)**
   alpha helix (α helix)   **disulfide bond**
   beta-pleated sheet (β–pleated sheet) **globular protein**
   tertiary (3°) structure   **fibrous protein**
   nonpolar interactions   **quaternary (4°) structure**
   B. Distinguish the levels of protein structure.

Copyright © 2014 Pearson Education, Inc.

C. Describe the attractive forces present as a protein folds into its three-dimensional shape.

10.4 Denaturation of Proteins

Learning Objectives

Upon completion of this material, a student should be able to do the following:
   A. Define the following key term:

   denaturation
   B. Define protein denaturation.
   C. List the causes of protein denaturation and the attractive forces affected.

10.5 Protein Functions

Learning Objectives

Upon completion of this material, a student should be able to do the following:
   A. Define the following key terms:

hormone	**prosthetic group**
receptors	**conformational change**
integral membrane protein	**antigen**

   B. Identify various functions of proteins.
   C. Provide examples of protein structure dictating protein function.

10.6 Enzymes – Life's Catalysts

Learning Objectives

Upon completion of this material, a student should be able to do the following:
   A. Define the following key terms:

active site	**coenzyme**
substrate	**enzyme-substrate complex (ES)**
substrate specificity	**lock-and-key model**
cofactor	**induced-fit model**

   B. Define active site and substrate.
   C. Distinguish the lock-and-key model from the induced-fit model.
   D. Discuss factors that lower the activation energy for an enzyme-catalyzed reaction.

10.7 Factors That Affect Enzyme Activity

Learning Objectives

Upon completion of this material, a student should be able to do the following:
   A. Define the following key terms:

activity	**reversible inhibition**
steady state	**competitive inhibitors**
pH optimum	**noncompetitive inhibition**
temperature optimum	**reversible inhibition**
irreversible inhibition	

   B. Describe how substrate concentration, pH, and temperature affect enzyme activity.
   C. Distinguish competitive, noncompetitive, and irreversible inhibition.

Copyright © 2014 Pearson Education, Inc.

Practice Test for Chapter 10

1. The side chain in the following amino acid is a:

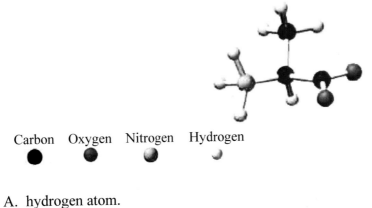

   Carbon Oxygen Nitrogen Hydrogen

   A. hydrogen atom.
   B. methyl group.
   C. carboxylate group.
   D. amino group.
   E. none of the above

2. The following amino acid could be represented as:

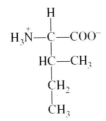

   A. isoleucine.
   B. Ile.
   C. I.
   D. all of the above

3. How many chiral centers are present in the following amino acid?

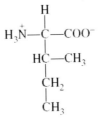

   A. none
   B. 1
   C. 2
   D. 3
   E. 5

Copyright © 2014 Pearson Education, Inc.

4. When amino acids react and form a peptide:
   A. a condensation reaction occurs.
   B. an amide bond forms.
   C. the —COO⁻ of one amino acid reacts with the —NH₃⁺ of another amino acid.
   D. both A and B
   E. A, B, and C

5. Consider the following structure.

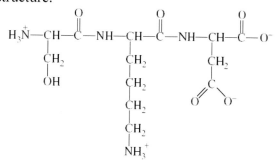

   This represents a _____ with_____ as the N-terminal amino acid.

   Fill in the blanks with the correct choice.
   A. tripeptide, serine
   B. tripeptide, aspartate
   C. dipeptide, Ser
   D. dipeptide, D

6. Consider the following structure.

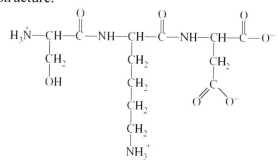

   Which of the following correctly describes this structure?
   A. contains three peptide bonds and all polar side chains
   B. contains three peptide bonds and all hydrophilic side chains
   C. contains two peptide bonds and all polar side chains
   D. contains two peptide bonds and all hydrophobic side chains

Copyright © 2014 Pearson Education, Inc.

7. The following peptide could be named as:

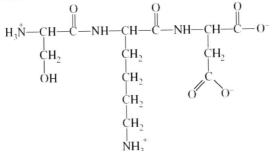

A. Ser—Lys—Asp.
B. S—L—A.
C. Asp—Lys—Ser.
D. D—R—S.
E. either A or B
F. either C and D

8. Consider the following.

Asp—Lys—Ser—Asp—Arg—Val—Tyr—Ile—His—Pro—Phe—Ser—Lys—Asp

What level of protein structure is represented?
A. primary
B. secondary
C. tertiary
D. quaternary

9. When the following two peptides interact via hydrogen bonding, what level of protein structure is formed?

Asp—Lys—Ser—Asp—Arg—Val—Tyr—Ile—His—Pro—Phe—Ser—Lys—Asp

Tyr—Gln—Val—Asp—Ser—Lys—Thr—Pro—Ala—Val—Ile—Leu—Gln—Arg

A. primary
B. secondary
C. tertiary
D. quaternary

Copyright © 2014 Pearson Education, Inc.

10. What level of protein structure is depicted in the following image?

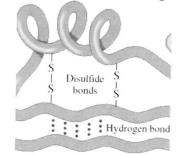

A. primary
B. secondary
C. tertiary
D. quaternary

11. Which level of protein structure does not usually involve hydrogen bonding?
A. primary
B. secondary
C. tertiary
D. quaternary

12. Consider the following description of protein structure:

   "Open extended zigzag with amino acid side chains pointing upward and downward"

   This description corresponds to a:
   A. secondary structure called a β-pleated sheet.
   B. tertiary structure called a β-pleated sheet.
   C. secondary structure called an α helix.
   D. tertiary structure called an α helix.

13. How many different tripeptides could be produced containing one glycine, one cytosine, and one glutamine?
   A. 1
   B. 2
   C. 3
   D. 4
   E. 6
   F. 8

Copyright © 2014 Pearson Education, Inc.

14. The amino acid shown would most likely be involved in what type of interaction in the tertiary structure of protein?

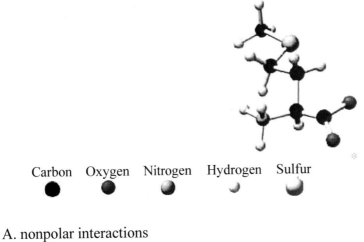

Carbon   Oxygen   Nitrogen   Hydrogen   Sulfur

A. nonpolar interactions
B. polar interactions
C. salt bridges
D. disulfide bridges

15. What type(s) of interaction(s) could occur between the side chains of the following pairs of amino acids?

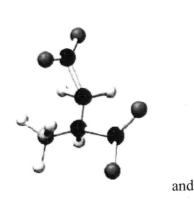   and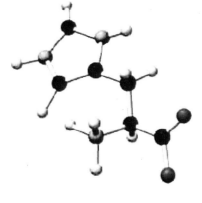

Carbon   Oxygen   Nitrogen   Hydrogen

A. nonpolar interactions
B. polar interactions
C. salt bridges
D. disulfide bonds
E. all of the above

16. The stabilizing force in the secondary structure of proteins is:
A. nonpolar interactions.
B. polar interactions.
C. salt bridges.
D. disulfide bridges.
E. hydrogen bonding.

Copyright © 2014 Pearson Education, Inc.

17. Consider the following image.

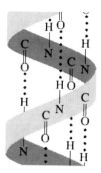

This excerpt represents a:
A. secondary protein structure stabilized by hydrogen bonds.
B. primary protein structure stabilized by hydrogen bonds.
C. secondary protein structure stabilized by salt bridges.
D. tertiary protein structure stabilized by hydrogen bonds.

18. The following represents an excerpt from a polypeptide chain:

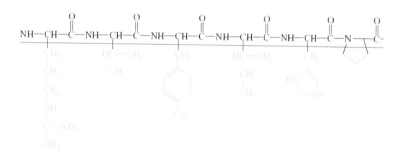

Which of the following correctly describes this image?

A. The underlined portion represents the protein backbone.
B. This is a representation of the primary structure of the protein.
C. The gray structures represent the protein's side chains.
D. All of the above are correct for this image.

19. When two —SH groups are brought close together when a protein folds:
A. reduction can occur, forming a disulfide bond.
B. oxidation can occur, forming an —S—S— link, which stabilizes the tertiary structure of the protein.
C. oxidation can occur, forming a salt bridge.
D. reduction can occur, which stabilizes the quaternary structure of the protein.

20. Which of the following amino acids would be more likely to be on the interior of a protein after it has folded into its tertiary structure?
A. lysine
B. aspartate
C. tyrosine
D. phenylalanine

Copyright © 2014 Pearson Education, Inc.

21. Which of the following is most likely to disrupt the hydrogen bonding in proteins?
    A. application of heat
    B. addition of an acid
    C. addition of a base
    D. presence of organic compounds

22. Which of the following denaturing agents would most likely disrupt the stabilizing force represented here?

    A. addition of an acid
    B. heavy metals
    C. mechanical agitation
    D. both B and C

23. Consider the following depiction of a protein.

    This image could represent:
    A. hemoglobin.
    B. an antibody.
    C. collagen.
    D. a glucose transporter.

24. Consider the following depiction of the reaction between glucose and adenosine triphosphate.

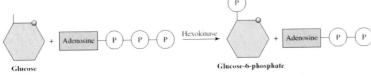

    In this reaction:
    A. glucose-6-phosphate is the ES.
    B. hexokinase is the enzyme.
    C. adenosine triphosphate is a cofactor.
    D. all of the above

Copyright © 2014 Pearson Education, Inc.

25. Which of the following models of enzyme activity could be used to explain an enzyme that has more than one substrate?

A.

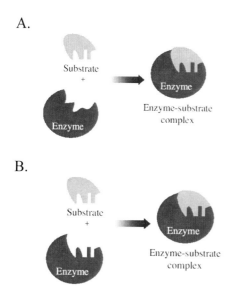

B.

C. Either one could be used.
D. Neither one because enzymes are specific.

26. Carboxypeptidase A catalyzes the hydrolysis of peptide bonds. Zn^{2+} is needed in this hydrolysis. In this reaction:
 A. carboxypeptidase A is a coenzyme and Zn^{2+} is a cofactor.
 B. carboxypeptidase A is the enzyme and Zn^{2+} is a coenzyme.
 C. carboxypeptidase A is a cofactor and Zn^{2+} is a coenzyme.
 D. carboxypeptidase A is the enzyme and Zn^{2+} is a cofactor.

27. The following image shows glucose and the enzyme hexokinase (the large molecule). Based on this diagram:

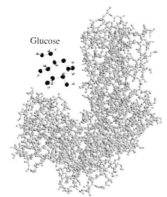

 A. the crevice or pocket is the active site of hexokinase.
 B. the ES has not yet formed.
 C. the crevice or pocket might be lined with polar amino acids.
 D. all of the above

Copyright © 2014 Pearson Education, Inc.

28. Which of the following is **not** a factor in lowering the activation energy in the formation of ES?
    A. heat of reaction
    B. bond energy
    C. orientation
    D. proximity

29. Sucrase catalyzes the breakdown of sucrose to glucose and fructose in the small intestine with an optimum pH of 6.3. Which of the following will decrease the rate of catalysis?
    A. increase the pH
    B. decrease the H_3O^+ concentration
    C. increase the temperature to 65 °C
    D. all of the above

30. The structure on the left has been shown to inhibit the reaction of the substrate on the right.

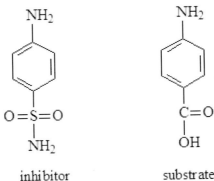

inhibitor   substrate

This is most likely an example of:
    A. reversible inhibition.
    B. irreversible inhibition.
    C. competitive inhibition.
    D. noncompetitive inhibition.
    E. A and C
    F. A and D

Copyright © 2014 Pearson Education, Inc.

Answers

1. B **2.** D **3.** C **4.** E **5.** A **6.** C **7.** A **8.** A **9.** B **10.** C
11. A **12.** A **13.** E **14.** A **15.** C **16.** E **17.** A **18.** D **19.** B **20.** D
21. A **22.** B **23.** C **24.** B **25.** A **26.** D **27.** D **28.** A **29.** D **30.** E

Copyright © 2014 Pearson Education, Inc.

Chapter 10 – Solutions to Odd-Numbered Problems

Practice Problems

10.1   a.   b.

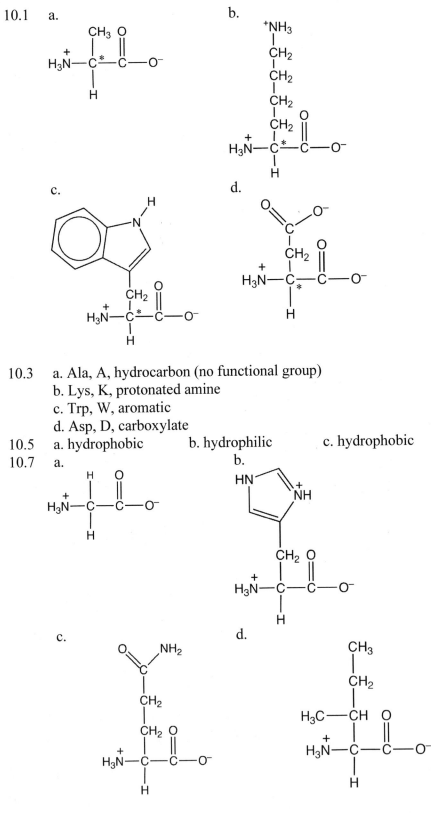

10.3   a. Ala, A, hydrocarbon (no functional group)
        b. Lys, K, protonated amine
        c. Trp, W, aromatic
        d. Asp, D, carboxylate

10.5   a. hydrophobic   b. hydrophilic   c. hydrophobic   d. hydrophilic

10.7   a.

10.9 a. glycine   b. histidine   c. glutamine   d. isoleucine
10.10 a. proline   b. asparagine   c. valine   d. tyrosine
10.11

a.

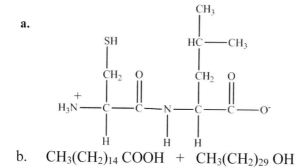

b. $CH_3(CH_2)_{14} COOH$ + $CH_3(CH_2)_{29} OH$

10.13 a.

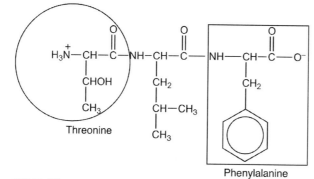

b.   TLF, Thr–Leu–Phe

10.15 & 10.17

a.

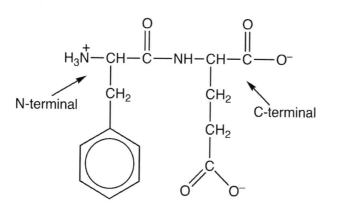

Copyright © 2014 Pearson Education, Inc.

b.

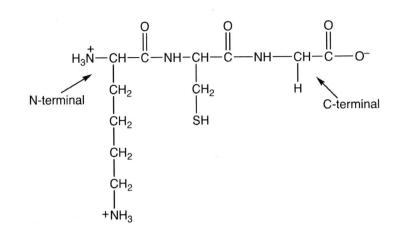

c.

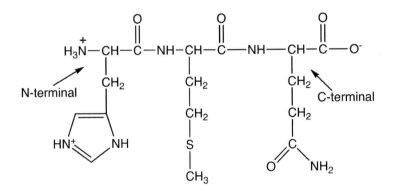

d.

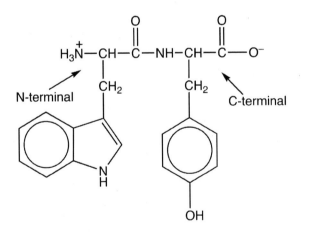

10.19 covalent bonding; peptide bond

10.21 hydrogen bonding

10.23 An α-helix is a coiled structure with the amino acid side chains protruding outward from the helix. The β-pleated sheet is an open, extended, zig-zag structure with the side chains of the amino acids oriented above and below the sheet.

10.25 a. salt bridge   b. London force (nonpolar interaction)
      c. hydrogen bonding   d. hydrogen bonding

10.27 a. quaternary   b. secondary   c. tertiary   d. primary

Copyright © 2014 Pearson Education, Inc.

10.29   a. ionic and hydrogen bonding; secondary, tertiary, and quaternary structure
        b. hydrogen bonding and London forces; secondary, tertiary, and quaternary structure
10.31   collagen, structural connector
        hemoglobin, oxygen transporter
        antibody, bind foreign substances in body
        casein, storage protein
10.33   After a meal, the glucose transporter is active because glucose levels are high in the bloodstream
        and insulin is released. Upon waking up in the morning, the glucose transporter is less active
        because glucose and insulin levels in the bloodstream are low.
10.35   tertiary
10.37   The induced-fit model. When glucose is fit into the active site, the enzyme undergoes a
        conformational change, closing around the substrate.
10. 39 Because the phosphates on ATP have negative charges and Mg^{2+} has positive charges,
        the attraction is ionic.
10.41   a. The enzyme activity will decrease if the temperature is raised above the optimum temperature.
        b. The activity will be lowered if the pH is changed.
10.43   a. no effect b. rate decreases

Additional Problems

10.45   carboxylate and protonated amine
10.47   a. tyrosine, Tyr
        b. proline, Pro
        c. cysteine, Cys
10.49   isoleucine, Ile, and threonine, Thr
10.51   Many possibilities, among them are: rice and beans, peas and corn, oatmeal and peas, peas and
        rice
10.53   a. 6
        b. Gly-Pro-Lys, Gly-Lys-Pro, Pro-Gly-Lys, Pro-Lys-Gly, Lys-Pro-Gly, Lys-Gly-Pro
        c.

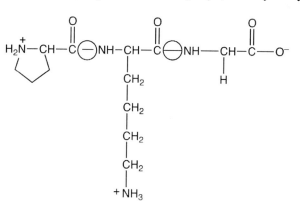

Copyright © 2014 Pearson Education, Inc.

10.55 a.

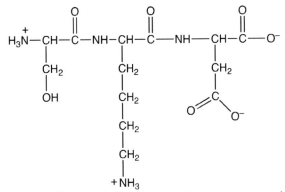

b. This segment would be on the surface in the aqueous environment. The side chains are hydrophilic and would interact with the water in the surrounding environment.

10.57 a. Primary structures are held together by covalent bonding and secondary structures are held together by hydrogen bonds forces.

b. Complete proteins are those that contain all the essential amino acids and incomplete proteins do not.

c. Fibrous proteins have elongated structures and globular proteins have roughly spherical structures. Globular proteins are water soluble but fibrous proteins are not.

10.59 a. quaternary structure

b. secondary structure

c. primary structure

d. tertiary and quaternary structure

10.61 Collagen's secondary structure is a helix of three helices woven together. This structure is called a triple helix.

10.63 a. hydrogen bonds and nonpolar attractions; secondary, tertiary, and quaternary

b. hydrogen bonds and salt bridges; secondary, tertiary, and quaternary

c. hydrogen bonds and nonpolar attractions; secondary, tertiary, and quaternary

10.65 disulfide bonds, thioglycolate

10.67 a. proline

b. An OH group is added to the side chain to form hydroxyproline.

c. The OH on the side chain allows for the formation of more hydrogen bonds between side chains, which increases the strength of the collagen.

10.69 The middle portion of these proteins span the nonpolar portion of the membrane. This surface interacts with the nonpolar environment and, therefore, must be nonpolar itself.

10.71 The active site is the location on an enzyme where catalysis occurs.

10.73 a. (1) active site

b. (3) induced-fit model

c. (2) lock-and-key model

10.75 Because trypsin can have more than one substrate, its action is better described by the induced-fit model.

10.77 Since thiamine is an organic molecule involved in enzymatic reactions, but is not itself an enzyme, then it must be a coenzyme.

10.79 a. requires a cofactor

b. describes a simple enzyme

c. requires a coenzyme

10.81 a. (c) both

b. (b) serve to hold the substrate

c. (b) serve to hold the substrate

Copyright © 2014 Pearson Education, Inc.

        d. (c) both
10.83 a. The rate would increase due to an increase of the probability of collision until a steady state is reached.

     b. The rate would decrease due to protein denaturation.

     c. The optimum temperature for trypsin is near body temp, 37 °C. The rate would decrease due to slower movement of molecules.

10.85 The products do not fit snugly into the active site, so they are released.

10.87 An irreversible inhibitor forms bonds to the enzyme and permanently inactivates it. A reversible inhibitor temporarily inactivates the enzyme, but when the inhibitor is removed, the enzyme regains its activity.

10.89 If the amount of enzyme was increased in a reaction occurring at a steady state, the maximum rate of the reaction would increase since more enzyme is present to convert substrate into product.

10.91 Cadmium would be a noncompetitive inhibitor because increasing the substrate and cofactor has no effect on the rate. This implies that the cadmium is binding to another site on the enzyme.

10.93 When designing an inhibitor, usually the substrate of an enzyme is known even if the structure of the enzyme is not. It is easier to design a molecule that resembles the substrate (competitive) than to find an inhibitor that binds to a second site on an enzyme (noncompetitive).

Challenge Problems

10.95

10.97 Both have nonpolar interiors and polar surfaces. On a micelle, the polar heads of the fatty acid salts face outwards into the aqueous environment, similar to the polar amino acid side chains on the globular protein. The nonpolar tails of the fatty acid salts gather together in the interior of the micelle just as the nonpolar amino acid side chains gather in the interior of the globular protein.

10.99 a. The polar amino acids are likely involved in catalysis.

     b. The nonpolar amino acids are likely involved in aligning the substrate correctly.

     c. The substrate is likely nonpolar with a small polar portion.

Copyright © 2014 Pearson Education, Inc.

Nucleic Acids — Big Molecules with a Big Role

11.1 Components of Nucleic Acids

Learning Objectives

Upon completion of this material, a student should be able to do the following:

    A. Define the following key terms:

        DNA   **RNA**

        genome   **nucleotide**

        gene   **nucleoside**

    B. Identify the five nitrogenous bases found in nucleic acids.

    C. Distinguish the bases ribose and deoxyribose.

    D. Write nucleosides and nucleotides given their component parts.

11.2 Nucleic Acid Formation

Learning Objectives

Upon completion of this material, a student should be able to do the following:

    A. Define the following key terms:

        nucleic acid   **phosphodiester bond**

    B. Write the product of a condensation of nucleotides.

    C. Abbreviate a nucleic acid using one-letter base coding.

11.3 DNA

Learning Objectives

Upon completion of this material, a student should be able to do the following:

    A. Define the following key terms:

        double helix   **supercoiling**

        complementary base pair   **chromosome**

    B. Characterize the structural features of DNA.

    C. Write the complementary base pairs for a single strand of DNA.

11.4 RNA and Protein Synthesis

Learning Objectives

Upon completion of this material, a student should be able to do the following:

    A. Define the following key terms:

        transcription   **translation**

Copyright © 2014 Pearson Education, Inc.

mRNA	**tRNA**
RNA polymerase	**anticodon**
rRNA	

B. List the types of RNA and their role in protein synthesis.

C. Translate a DNA strand into its complementary mRNA.

11.5 Putting It Together: The Genetic Code and Protein Synthesis

Learning Objectives

Upon completion of this material, a student should be able to do the following:

    A. Define the following key terms:

codon	**genetic code**

B. Distinguish transcription from translation.

C. Translate a mRNA sequence into a protein sequence using the genetic code.

11.6 Genetic Mutations

Learning Objectives

Upon completion of this material, a student should be able to do the following:

    A. Define the following key terms:

mutation	**carcinogen**
silent mutation	**somatic cell**
spontaneous mutation	**germ cell**
mutagen	**genetic disease**

B. Define genetic mutation.

C. Determine changes in protein sequence if a mRNA sequence is mutated.

11.7 Viruses

Learning Objectives

Upon completion of this material, a student should be able to do the following:

    A. Define the following key terms:

host	**envelope**
capsid	**retrovirus**

B. List the differences between a virus and a cell.

C. List the structural components of a virus.

D. Describe how a virus infects a cell.

11.8 Recombinant DNA Technology

Learning Objectives

Upon completion of this material, a student should be able to do the following:

    A. Define the following key terms:

recombinant DNA	**plasmid**
restriction enzyme	**expression**
vector	**clone**

B. Apply knowledge of nucleic acid structure to DNA technology.

Copyright © 2014 Pearson Education, Inc.

Practice Test for Chapter 11

1. The following substance is classified as:

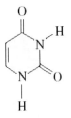

   A. a purine.
   B. a pyrimidine.
   C. ribose.
   D. deoxyribose.

2. The following substance is classified as:

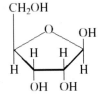

   A. a purine.
   B. a pyrimidine.
   C. ribose.
   D. deoxyribose.

3. If a condensation reaction occurred between the following two substances, the type of product formed would be:

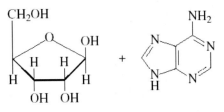

   A. DNA.
   B. a nucleotide.
   C. a nucleoside.
   D. RNA.

Copyright © 2014 Pearson Education, Inc.

4. What is the name of the product of the following reaction?

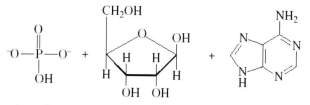

A. adenosine-5'-monophosphate
B. adenosine nucleotide
C. adenosine phosphate
D. AMP
E. A or D

5. Where is the following substance usually found?

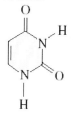

A. RNA
B. DNA
C. both DNA and RNA
D. neither DNA nor RNA

6. The difference between CMP and dCMP is found in which component of their molecular structure?
A. sugar
B. phosphate
C. nitrogenous base
D. the bonds linking the components together

7. Consider the following structure.

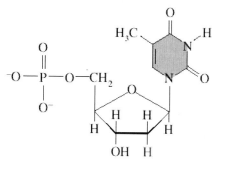

This represents:
A. an RNA nucleotide.
B. uridine-5'-monophosphate.
C. TMP.
D. dTMP.

Copyright © 2014 Pearson Education, Inc.

8. What type of bond connects the nucleotides in a nucleic acid?
   A. amide
   B. phosphodiester
   C. peptide
   D. ester

9. Consider the following structure.

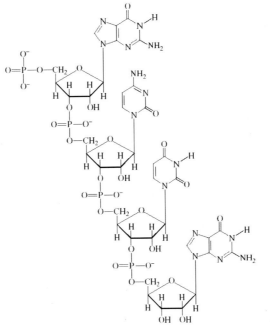

Which of the following is correct?
A. has pyrimidine bases at the 5' and 3' ends
B. contains 4 phosphodiester bonds
C. contains 4 nucleotides
D. could be found in DNA
E. All of the above are correct.

Copyright © 2014 Pearson Education, Inc.

10. What is the abbreviation for the following structure?

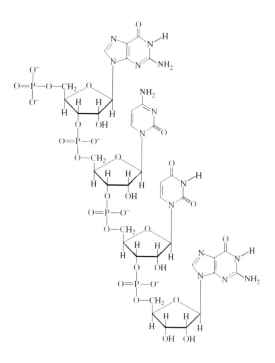

    A. 5'GCUG3'
    B. 5'GUCG3'
    C. 5'GUTG3'
    D. 5'GAUG3'

11. The nucleic acid analog of the N-terminus of a protein is:
    A. a nucleotide.
    B. the 5' end.
    C. the 3' end.
    D. a phosphodiester bond.

12. The double helix structure of DNA is stabilized by:
    A. phosphodiester bonds.
    B. covalent bonds.
    C. hydrogen bonds.
    D. disulfide bridges.

13. Which of the following is **not** a complementary base pair found in DNA?
    A. A and T
    B. G and C
    C. A and U
    D. All are complementary DNA base pairs.

Copyright © 2014 Pearson Education, Inc.

14. What is the complementary DNA segment 3' to 5' for the segment shown below?

5'GGCTTA3'

A. 3'ATTCGG5'
B. 3'GGCTTA5'
C. 3'TAAGCC5'
D. 3'CCGAAT5'

15. Which of the following constitutes a difference between DNA and RNA?
   A. sugar present
   B. strand structure
   C. size
   D. complementary base of adenine
   E. All of the above are differences.

16. Which base in mRNA complements the base represented as "A" in DNA?
   A. T
   B. U
   C. G
   D. C

17. The sequence of bases in a DNA template is:

5'ATCGAT3'

What is the corresponding mRNA that is produced from this DNA?
   A. 3'TAGCTA5'
   B. 3'TAGCTA5'
   C. 3'UAGCUA5'
   D. 3'AUCGAU5'

18. Which type of RNA provides the amino acids for the growing peptide chain?
   A. tRNA
   B. mRNA
   C. rRNA
   D. none of the above

19. The synthesis of a protein based on a DNA template is called:
   A. translocation.
   B. transcription.
   C. translation.
   D. transmutation.

Copyright © 2014 Pearson Education, Inc.

20. What do the following mRNA codons have in common?

UAA   UAG   UGA

A. stop protein synthesis
B. start protein synthesis
C. code for the same amino acid
D. have nothing in common

21. What is the sequence of amino acids coded by the following codons in mRNA?

5' CCA|GUC|AAA|GCC 3'

A. Ala—Lys—Val—Pro
B. Pro—Lys—Val—Ala
C. Pro—Val—Lys—Ala
D. Ala—Val—Lys—Pro

22. The conversion of the information on the codons in mRNA to a peptide chain occurs:
A. in the nucleus of the cell.
B. on the nuclear membrane.
C. in the cytoplasm.
D. in the ribosome.

23. Consider the following representation of tRNA.

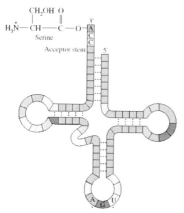

This tRNA is a complement to what codon on mRNA?
A. 5'AGU3'
B. 5'UCA3'
C. 5'TCA3'
D. 5'ACU3'

24. Which of the following represents the correct sequence of events in protein synthesis?
A. transcription, activation, translation, termination
B. transcription, activation, termination, translation
C. translation, activation, transcription, termination
D. activation, transcription, translation, termination

Copyright © 2014 Pearson Education, Inc.

25. Consider the following segment of DNA.

3' GGG|TTA|CAC|ATT 5'

What amino acids will be placed in the peptide chain from this segment?
A. Val—Asn—Pro
B. Pro—Asn—Val—Tyr
C. Pro—Asn—Val
D. Val—Leu—His—Ile

26. If a codon in mRNA was changed from UGU to UGC, this change could most specifically be called a:
A. mutagen.
B. mutation.
C. silent mutation.
D. carcinogen.

27. A normal DNA template produces the following mRNA sequence:

5'ACU|UAC|CGG3'

If a change in the DNA template changes the mRNA sequence to:

5'ACU|UAG|CGG3'

How would this change affect the protein produced?
A. silent mutation
B. mutation affecting protein structure/function
C. no effect at all
D. cannot predict with the given information

28. Which of the following is **not** a property of all viruses?
A. utilize the ribosomes of the infected cells
B. contain nucleic acids enclosed in a capsid
C. can infect any cell type
D. consist of small particles with 3–200 genes
E. All of the above are properties of all viruses.

29. Which of the following correctly pertains to the HIV-1 virus and current therapies?
A. HIV-1 is a retrovirus.
B. Nucleoside analogs are competitive inhibitors.
C. Protease inhibitors halt transcription.
D. Current research involves the use of mutagens.

30. In recombinant DNA technology, the synthesis of a protein cloned into a second organism is called a(n):
A. vector.
B. somatic cell.
C. expression.
D. restriction enzyme.

Answers

1. B   2. C   3. C   4. E   5. A   6. A   7. D   8. B   9. C   10. A
11. B   12. C   13. C   14. D   15. E   16. B   17. C   18. A   19. C   20. A
21. C   22. D   23. B   24. A   25. C   26. C   27. B   28. E   29. A   30. C

Copyright © 2014 Pearson Education, Inc.

Chapter 11 – Solutions to Odd-Numbered Problems

Practice Problems

11.1   a. pyrimidine   b. pyrimidine
11.3   a. DNA   b. both DNA and RNA
11.5   deoxyadenosine-5'-monophosphate (dAMP), deoxythymidine-5'-monophosphate (dTMP),
deoxycytidine-5'-monophosphate (dCMP), deoxyguanosine-5'-monophosphate (dGMP)
11.7   a.

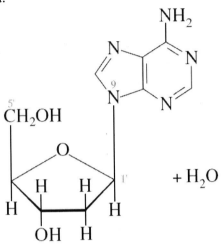

$+ H_2O$

b.

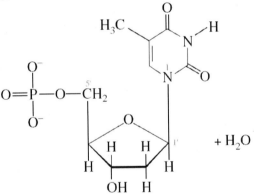

$+ H_2O$

11.9   The nucleotides in nucleic acids are held together by phosphodiester bonds between the 3'-OH of
a sugar (ribose or deoxyribose) and a phosphate group on the 5'-carbon of another sugar.

Copyright © 2014 Pearson Education, Inc.

11.11

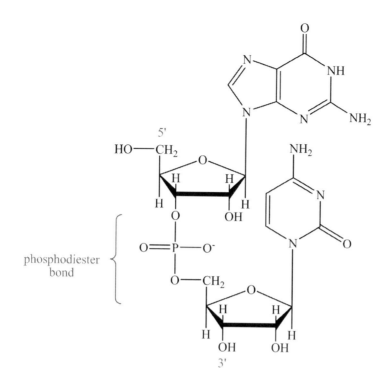

phosphodiester bond

11.13 The two DNA strands are held together by hydrogen bonds between the bases in each strand.

11.15 a. 3'TTTT5'

b. 3'GGGGAAAA5'

c. 3'TGTAACC5'

d. 3'ACACTTGG5'

11.17

	Protein	Nucleic Acid
Primary structure	Sequence of amino acids	Sequence of nucleotides
Secondary structure	Local hydrogen bonding between backbone atoms.	Hydrogen bonding between bases in nucleic acid strands
Tertiary structure	Folding of backbone associating amino acids far away in sequence.	Supercoiling of DNA into tight, compact structures like chromosomes.

11.19 Messenger RNA (mRNA): It contains a base sequence complementary to the DNA strand that encodes for proteins. It copies DNA and then moves from the nucleus to the ribosome.
Ribosomal RNA (rRNA): The structural component of the ribosome where proteins are manufactured in the cell.
Transfer RNA (tRNA): It is used in the ribosome to place a specific amino acid in the growing protein chain.

11.21 In transcription, the sequence of nucleotides on one strand of a DNA template (the double helix is temporarily unwound) is used to produce a complementary mRNA copy using the enzyme RNA polymerase.

11.23 3'AUGCCGUUCGAU5'

11.25 There are 20 amino acids, so there must be at least 20 different tRNAs.

11.27 a. Lys–Pro–Leu–Ala

b. Pro–Arg–Ser–Pro–Stop

Copyright © 2014 Pearson Education, Inc.

c. Met–His–Lys–Glu–Val–Leu

11.29 Translocation is the movement of the second tRNA to the spot vacated by the first tRNA, allowing the next tRNA to bind to the ribosome. The amino acid is added to the growing peptide.

11.31 a. 5′ACA|CCC|CAA|UAA3′

b. 3′UGU|GGG|GUU|AUU5′

c. three-letter code: Thr–Pro–Gln–Stop, one-letter code: TPQ–Stop

11.33 A mutagen is an environmental agent that produces a mutation (change in base sequence) in DNA.

11.35 a. three-letter code: Leu–Lys–Arg–Val, one-letter code: LKRV

b. three-letter code: Ile–Lys–Arg–Val, one-letter code: IKRV

Changing leucine to isoleucine, which are both nonpolar, will probably have no effect on the protein function.

c. Nothing. Both codons code for arginine.

d. Leucine will be the last amino acid incorporated into the growing protein chain because UAA is a stop codon.

e. If an A is added to the beginning of the chain, the new mRNA would be: 5′ACU|UAA|ACG|AGU3′.

The amino acid sequence would be: Thr–Stop (T-Stop)

f. If C is removed from the beginning of the chain, the new mRNA would be: 5′UUA|AAC|GAG3′.

The new amino acid sequence would be: Leu–Asn–Glu (LNE).

11.37 Viruses cannot replicate themselves without a host cell. They do not have the cell machinery (enzymes, organelles, etc.).

11.39 A vaccine allows the body to mount an immune response, by producing antibodies, to a less active form of the virus. Then, if an active form of the virus is encountered later, the body can fight against it more effectively.

11.41 Protease inhibitors inactivate an HIV-1 specific protease that is needed to process viral proteins.

11.43 Gene expression is producing a protein from a DNA gene sequence (usually of a non-native gene).

11.45 A vector is a transporter for donor DNA. It enables the incorporation of the donor DNA into the organism DNA.

Additional Problems

11.47 a. pyrimidine   b. purine   c. pyrimidine
      d. pyrimidine   e. purine

11.49 a. cytosine, ribose
      b. adenine, deoxyribose
      c. guanine, deoxyribose
      d. uracil, ribose

11.51 Thymine contains a methyl ($-CH_3$) group at carbon-5 that uracil lacks.

Copyright © 2014 Pearson Education, Inc.

11.53

	Protein	Nucleic Acid
Repeating unit	amino acid	nucleotide
Backbone repeat	N–Cα–C–N–Cα–C–etc.	sugar-phosphate-sugar-phosphate-etc.
One letter abbreviation	name of amino acid	name of base
Free left end	amino or N terminus	5′ end
Free right end	carboxy or C terminus	3′ end

11.55 a. 3′AATGCCTAGGCG5′

b. 3′TATCGGGAATGACC5′

c. 3′CCGGATGGAATTGCTGC5′

11.57 a. mRNA   b. tRNA   c. rRNA

11.59 a. GUU, GUC, GUA, GUG

b. CCU, CCC, CCA, CCG

c. CAU, CAC

11.61 a. leucine   b. arginine   c. isoleucine

11.63 three-letter code for Met-enkephalin: Tyr–Gly–Gly–Phe–Met, one-letter code: YGGFM

11.65 a. CAC   b. GGG   c. CUU

11.67 a. Both alanine and leucine have small, nonpolar R-groups. Both amino acids are found in similar environments in proteins, so this substitution is unlikely to affect protein structure or function.

b. If a serine codon (UCA) is replaced with a stop codon (UAA), the serine and remaining amino acids will not be added to the growing polypeptide chain. This can affect both the structure and function of the protein.

11.69 a. The development of nucleoside analogs that can be incorporated into viral DNA to stop reverse transcription is viable.

b. Integrase is a viral-only enzyme, so this would be a viable method for inactivating the virus.

c. Since RNA polymerase is used to make all proteins in the cell, this would not be a viable method to inactivate the virus.

11.71 Entry-point inhibitors block sites on the outside of a host cell where a virus would attach and eventually enter. If the virus cannot enter the cell, the virus cannot replicate.

11.73 In gene cloning, a copy of a single gene is created. In organism cloning, many genes are copied and an exact copy of an organism is created.

Challenge Problems

11.75 3 nucleotides/amino acid + 1 start codon + 1 stop codon = 35 amino acids x 3 + 3 + 3 = 111 nucleotides

11.77 The more hydrogen bonds present, the more heat (higher temperature) must be applied to denature the DNA. Because strand 2 has more G–C base pairs with three hydrogen bonds each, it is predicted to have the higher denaturation temperature.

Food as Fuel —
A Metabolic Overview

12.1   How Metabolism Works

Learning Objectives
Upon completion of this material, a student should be able to do the following:
   A. Define the following key terms:

metabolism	**anabolism**
metabolic pathway	**cytoplasm**
metabolite	**cytosol**
catabolism	**mitochondrion**

   B. Distinguish catabolism from anabolism.
   C. Identify reactions as catabolic or anabolic.
   D. Name the parts of a cell associated with metabolism.

12.2 Metabolically Relevant Nucleotides

Learning Objectives
Upon completion of this material, a student should be able to do the following:
   A. Identify the metabolically relevant nucleotides.
   B. Distinguish the low-energy and high-energy forms of the relevant nucleotides.

12.3 Digestion— From Fuel Metabolism to Hydrolysis Products

Learning Objectives
Upon completion of this material, a student should be able to do the following:
   A. Define the following key terms:

digestion	**chylomicrons**
emulsification	

   B. Compare digestion of carbohydrates, lipids, and proteins.

12.4 Glycolysis— From Hydrolysis Products to Common Metabolites

Learning Objectives
Upon completion of this material, a student should be able to do the following:
   A. Define the following key terms:

gluconeogenesis	**oxidative decarboxylation**
aerobic	**fermentation**
anaerobic	

Copyright © 2014 Pearson Education, Inc.

B. Follow a molecule of glucose through the ten reactions of glycolysis.
C. Discuss anaerobic and aerobic fates of pyruvate.
D. Contrast glycolysis for glucose and for fructose.

12.5 The Citric Acid Cycle — Central Processing

Learning Objectives
Upon completion of this material, a student should be able to do the following:
    A. Define the following key term:
        citric acid cycle
    B. Identify the reactions in the citric acid cycle.
    C. List the energy output of the citric acid cycle.

12.6 Electron Transport and Oxidative Phosphorylation

Learning Objectives
Upon completion of this material, a student should be able to do the following:
    A. Define the following key terms:

oxidative phosphorylation	**electrochemical gradient**
chemiosmotic model	**thermogenesis**

    B. Describe the function of each enzyme complex (I-IV) during electron transport.
    C. Discuss the function of coenzyme Q and cytochrome *c*.
    D. Describe the production of ATP at complex V using the chemiosmotic model.

12.7 ATP Production

Learning Objectives
Upon completion of this material, a student should be able to do the following:
    A. Convert the number of reduced nucleotides produced (NADH, $FADH_2$) to a corresponding number of ATP.
    B. Calculate the number of ATP produced during the oxidative catabolism of a molecule of glucose.

12.8 Other Fuel Choices

Learning Objectives
Upon completion of this material, a student should be able to do the following:
    A. Define the following key terms:

beta oxidation (β oxidation)	**ketosis**
fatty acyl CoA	**transamination**
ketone bodies	**urea cycle**

    B. Calculate the number of ATP produced from a saturated fatty acid undergoing β oxidation.
    C. Describe the metabolic pathways of β oxidation, transamination, and the urea cycle.
    D. Identify catabolic and anabolic pathways in the cell.

Practice Test for Chapter 12

1. Which of the following reactions would **not** be classified as catabolic?

   A.
   $$ADP + P_i + Energy \longrightarrow ATP$$

   B.
   $$Protein + H_2O \xrightarrow[Enzyme]{H^+} Amino\ acids$$

   C.
   $$Maltose + H_2O \xrightarrow{Maltase} Glucose + glucose$$

   D.
   $$C_6H_{12}O_6 + 6O_2 \longrightarrow 6CO_2 + 6H_2O$$

2. Consider the substances listed in the choices below. Which would be classified as a metabolite?
   A. lactose
   B. carbon dioxide
   C. pyruvate
   D. ATP
   E. All are metabolites.

3. Which of the following is considered to be the energy currency of the body?
   A. FAD
   B. ATP
   C. NADH
   D. Acetyl CoA

4. How is the oxidized form of nicotinamide adenine dinucleotide represented?
   A. $FADH_2$
   B. NADH
   C. FAD
   D. NAD^+
   E. ADP

5. Which of the following is a characteristic of acetyl CoA?
   A. contains a thioester functional group
   B. is the high energy form of coenzyme A
   C. exchanges energy when a C$-$S bond is hydrolyzed
   D. A and B
   E. A, B, and C

Copyright © 2014 Pearson Education, Inc.

6. What is the abbreviation used to designate the following substance?

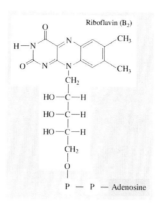

Riboflavin (B₂)

A. ATP
B. FAD
C. NAD$^+$
D. FADH$_2$
E. Co A
F. Acetyl CoA

7. The following reaction:

Lactose → Lactase → Galactose + Glucose

This reaction:
A. occurs in the stomach.
B. is classified as a condensation reaction.
C. produces products that can be transported in the bloodstream.
D. represents emulsification.

8. During the digestion of fats, which of the following transports the triglycerides from the stomach through the bloodstream?
A. chylomicron
B. bile salts
C. cholesterol
D. micelles

Copyright © 2014 Pearson Education, Inc.

9. Of the following reactions, which occurs in the stomach?

A.

Cholesterol + Free fatty acid → Cholesterol ester

B.

H₂C—O—
|
HC—O—   Pancreatic
| lipase
H₂C—O—   →

H₂C—OH
|
HC—O—
|
H₂C—OH

C.

Protein → Denatured protein

D.

Sucrose → Glucose + Fructose
     Sucrase

10. The end product of the digestion of starch is:
    A. cholesterol.
    B. amino acids.
    C. sucrose.
    D. glucose.

11. The following reaction represents the net chemical reaction of:

D-Glucose + $2NAD^+ + 2ADP + 2P_i$ → $2CH_3—C—C—O^- + 2NADH + 2H^+ + 2ATP$ (Pyruvate)

    A. glycolysis.
    B. the citric acid cycle.
    C. electron transport.
    D. oxidative phosphorylation.
    E. all of the above

Copyright © 2014 Pearson Education, Inc.

12. In terms of **high energy molecules**, what is the net output from two molecules of glucose during glycolysis?
    A. 2 ATP
    B. 2 ATP, 2 NADH
    C. 4 ATP, 2 NADH
    D. 4 ATP, 4 NADH
    E. 6 ATP, 6 NADH

13. Which of the following is produced during the aerobic oxidation of pyruvate?
    A. lactate
    B. acetyl CoA
    C. ethanol
    D. All are produced during aerobic oxidation.

14. Fructose can enter into glycolysis:
    A. as fructose-6-phosphate.
    B. as dihydroxyacetone phosphate.
    C. as glyceraldehyde-3-phosphate.
    D. after the regulatory step.
    E. all of the above

15. After 3 molecules of acetyl CoA move through the citric acid cycle:
    A. three molecules of NADH are produced.
    B. six molecules of $FADH_2$ are produced.
    C. six molecules of CO_2 are produced.
    D. six molecules of GTP are produced.

16. In order for the citric acid cycle to run, which of the following is needed?
    A. pyruvate and oxaloacetate
    B. acetyl CoA and pyruvate
    C. acetyl CoA and citrate
    D. oxaloacetate and acetyl CoA

17. What do the following two steps in the citric acid cycle have in common?

    Reaction 3: isocitrate $\longrightarrow$ α-ketoglutarate

    Reaction 8: malate $\longrightarrow$ oxaloacetate

    A. reduction of NAD^+
    B. production of CO_2
    C. decarboxylation
    D. isomerization
    E. reduction of FAD

18. Which of the following reaction types is not involved in the citric acid cycle?
    A. hydrogenation
    B. hydrolysis
    C. dehydrogenation
    D. isomerization

Copyright © 2014 Pearson Education, Inc.

19. All of the following are involved in electron transport **except**:
    A. cytochrome c.
    B. QH$_2$.
    C. ADP.
    D. FADH$_2$.

20. The process of producing energy from the oxidation of reduced nucleotides is termed:
    A. electron transport.
    B. oxidative phosphorylation.
    C. oxidative decarboxylation.
    D. dehydrogenation.

21. In the chemiosmotic model, which of the following drives the following reaction?

$$\text{ADP} + \text{P}_\text{i} + \text{energy} \longrightarrow \text{ATP}$$

    A. transfer of electrons from complex I to complex II
    B. release of H$^+$ during the oxidation of NADH and FADH$_2$
    C. reduction of O$_2$ to form water
    D. movement of H$^+$ through complex V

22. Electron transport occurs in the:
    A. cytosol.
    B. mitochondrial matrix.
    C. inner membrane of the mitochondria.
    D. cell membrane.

23. NADH enters electron transport at:
    A. complex I.
    B. complex II.
    C. complex III.
    D. complex IV.
    E. complex V.

24. For the oxidation of one molecule of glucose, which of the following produces the largest amount of ATP?
    A. glycolysis
    B. oxidation of pyruvate
    C. citric acid cycle
    D. All produce approximately equal amounts.

Copyright © 2014 Pearson Education, Inc.

25. The net production of ATP from NADH is greater than that from FADH$_2$ because:
   A. it takes two hydrogen atoms to reduce FAD.
   B. NADH enters the electron transport chain earlier than FADH$_2$.
   C. NADH pumps 6 protons into the inner mitochondrial membrane space while FADH$_2$ pumps 4 protons.
   D. A and C
   E. None of the above, FADH$_2$ produces more ATP per molecule.

26. Which of the following conversions directly produces the highest yield of ATP as a product of the reaction as written?

   A. glucose $\longrightarrow$ 2 pyruvate

   B. pyruvate $\longrightarrow$ lactate

   C. GTP $\longrightarrow$ GDP + P$_i$

   D. pyruvate $\longrightarrow$ acetyl CoA

27. Calculate the number of ATPs produced from the β oxidation of the following fatty acid entering β oxidation as an acyl CoA.

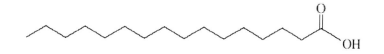

   A. 80 ATP
   B. 97.5 ATP
   C. 108 ATP
   D. 112 ATP

28. The following are the four steps involved in β oxidation. In which is the length of the fatty acyl CoA reduced by two carbon atoms?
   A. 1 - oxidation
   B. 2 - hydration
   C. 3 - oxidation
   D. 4 - removal of acetyl CoA
   E. None of the steps reduce the length by only two carbon atoms.

29. Which of the following characterizes the condition known as ketosis?
   A. accumulation of β-hydroxybutyrate, acetoacetate, and acetone in liver
   B. occurs in the presence of high concentration of carbohydrates
   C. low fat levels in the diet
   D. results in high blood pH

30. During metabolism, all but which of the following characterize amino acids?
   A. undergo transamination before excretion
   B. have the ability to directly generate ATP
   C. nitrogen content eventually excreted as urea
   D. can regenerate intermediates in the citric acid cycle

Copyright © 2014 Pearson Education, Inc.

Answers

1. A   2. C   3. B   4. D   5. E   6. B   7. C   8. A   9. C   10. D
11. A   12. D   13. B   14. E   15. C   16. D   17. A   18. A   19. C   20. B
21. D   22. C   23. A   24. C   25. B   26. A   27. C   28. D   29. A   30. B

Copyright © 2014 Pearson Education, Inc.

Chapter 12 – Solutions to Odd-Numbered Problems

Practice Problems

12.1   In metabolism, a catabolic reaction breaks apart molecules, releases energy, and is an oxidation.

12.3   a. catabolism   b. anabolism

12.5   hydrolysis

12.7   the mitochondrion

12.9   a. FAD/FADH2   b. Acetyl CoA

12.11 a. NADH   b. Acetyl CoA   c. FAD

12.13 a. starch   b. none   c. starch, oligosaccharides, disaccharides

12.15 Cholesterol is esterified and packaged into lipoproteins.

12.17 amino acids

12.19 d-glucose

12.21 4 ATP (net 2 ATP)

12.23 2 NADH and 2 ATP

12.25 NAD^+

12.27 ethanol, CO_2 and NAD^+

12.29 The main regulation point of glycolysis is phosphofructokinase, the enzyme involved in step 3. Because the products of the metabolism of fructose enter glycolysis at step 5, they bypass the main regulation point, which lead to the production of excess pyruvate and acetyl CoA that is not needed in the cells and is ultimately converted to fat.

12.31 acetyl CoA and oxaloacetate

12.33 isocitrate $\rightarrow$ α-ketoglutarate (reaction 3) and α-ketoglutarate $\rightarrow$ succinyl CoA (reaction 4)

12.35 isocitrate $\rightarrow$ α-ketoglutarate (reaction 3), α-ketoglutarate $\rightarrow$ succinyl CoA (reaction 4), and malate $\rightarrow$ oxaloacetate (reaction 8)

12.37 coenzyme Q

12.39 complex IV

12.41 As protons flow through ATP synthase, energy is released to produce ATP.

12.43 a. 2.5 ATP   b. 7 ATP   c. 5 ATP   d. 10 ATP

12.45 mitochondrial matrix

12.47 a. and b.

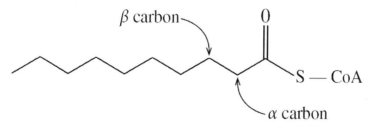

Copyright © 2014 Pearson Education, Inc.

c.

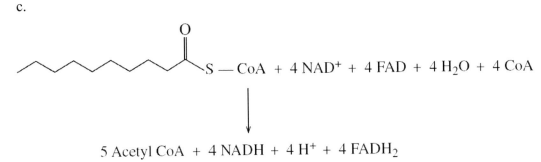

$$5 \text{ Acetyl CoA} + 4 \text{ NADH} + 4 \text{ H}^+ + 4 \text{ FADH}_2$$

12.49 Ketone bodies form when excess acetyl CoA results from the breakdown of large amounts of fat.

12.51 NH_4^+ is toxic if it accumulates in the body.

12.53 a. pyruvate   b. oxaloacetate, fumarate
       c. succinyl CoA   d. α-ketoglutarate

Additional Problems

12.55 a. cytosol
       b. mitochondrial matrix
       c. mitochondrial matrix

12.57 a. carbohydrate   b. fat   c. carbohydrate
       d. fat   e. protein

12.59 Lactose undergoes digestion in the small intestine to yield galactose and glucose.

12.61 a. produces ATP   b. neither   c. produces ATP

12.63 b. glucose-6-phosphate to fructose-6-phosphate

12.65 The runner's muscles may be switching over to anaerobic catabolism to keep ATP production going. The muscles produce lactate and H^+ during this process, causing soreness.

12.67 anaerobic (low oxygen) conditions

12.69 a. citrate and isocitrate
       b. α-ketoglutarate

       c. isocitrate $\rightarrow$ α-ketoglutarate (reaction 3), α-ketoglutarate $\rightarrow$ succinyl CoA (reaction 4), succinate $\rightarrow$ fumarate (reaction 6), and malate $\rightarrow$ oxaloacetate (reaction 8)

       d. isocitrate $\rightarrow$ α-ketoglutarate (reaction 3), and malate $\rightarrow$ oxaloacetate (reaction 8)

12.71 O_2 is used during electron transport. The coenzymes NADH and $FADH_2$ produced in the citric acid cycle are oxidized to FAD^+ and NAD during electron transport.

12.73 a. oxidized   b. reduced   c. reduced

12.75 matrix, intermembrane space

12.77 Energy released as protons flow down a concentration gradient through ATP synthase back to the matrix, energy is released and utilized for the synthesis of ATP.

12.79 The oxidation of glucose to pyruvate produces 7 ATP, 5 of which come from NADH; whereas 32 ATP are produced from the complete oxidation of glucose to CO_2 and H_2O.

12.81 a. 5   b. 4   c. 66 ATP

Copyright © 2014 Pearson Education, Inc.

Challenge Problems

12.83 Oxidation of pyruvate: pyruvate → acetyl CoA

Reaction 3 of citric acid cycle: isocitrate → α-ketoglutarate

Reaction 4 of the citric acid cycle: α-ketoglutarate → succinyl CoA

12.85 a. and b.

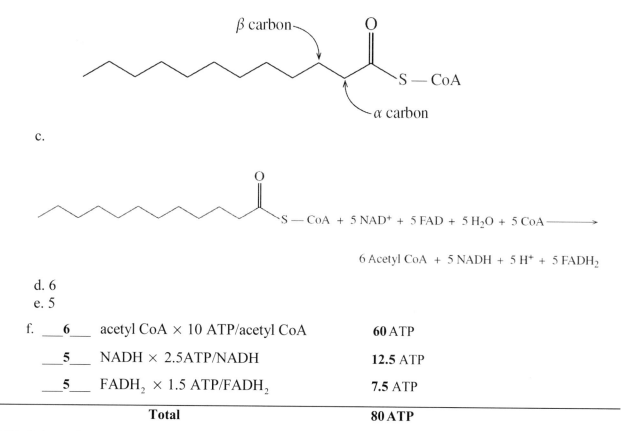

d. 6
e. 5

f. ___6___ acetyl CoA × 10 ATP/acetyl CoA   **60 ATP**

___5___ NADH × 2.5ATP/NADH   **12.5 ATP**

___5___ $FADH_2$ × 1.5 ATP/$FADH_2$ **7.5 ATP**

_____

                Total   **80 ATP**

12.87 Only oxaloacetate or pyruvate are starting points for gluconeogenesis. Acetyl CoA cannot be converted to pyruvate, so fatty acids cannot generate glucose.

Copyright © 2014 Pearson Education, Inc.